I0648567

Samuel George Shattock, William Wayne Babcock

An Atlas of the Bacteria Pathogenic in Man

Samuel George Shattock, William Wayne Babcock

An Atlas of the Bacteria Pathogenic in Man

ISBN/EAN: 9783744689298

Printed in Europe, USA, Canada, Australia, Japan

Cover: Foto ©berggeist007 / pixelio.de

More available books at **www.hansebooks.com**

AN ATLAS

OF THE

Bacteria Pathogenic in Man

WITH

DESCRIPTIONS OF THEIR MORPHOLOGY AND
MODES OF MICROSCOPIC EXAMINATION.

BY

SAMUEL G. SHATTOCK, F.R.C.S.,

Joint Lecturer on Pathology and Bacteriology, St. Thomas' Medical School,
London; Pathological Curator of the Museum of the
Royal College of Surgeons, London.

WITH AN INTRODUCTORY CHAPTER

ON

BACTERIOLOGY :

ITS PRACTICAL VALUE TO THE GENERAL PRACTITIONER.

BY

W. WAYNE BABCOCK, M.D.,

Pathologist to the Kensington Hospital for Women; Clinical Pathologist to the
Medical-Chirurgical Hospital; Demonstrator of Pathology and Bacteriology
in the Medico-Chirurgical College of Philadelphia.

SIXTEEN FULL-PAGE COLORED PLATES.

E. B. TREAT & CO.,
241–243 West 23d Street,
NEW YORK,
1899.   Price, $1.00

Publisher's Note.

This Atlas was originally published in the "International Medical Annual" in two sections, one in the 1898 issue, the other in the 1899 issue.

The perfection of the plates illustrating the appearance, under the microscope, of the Bacteria Pathogenic in the human subject was of such a high degree as to evoke the admiration of those familiar with Bacteriology. Then, too, the descriptive text was succinct and practical, and written for the use of physicians desirous of keeping up to date, but who have not the time nor inclination to master the elaborate text books on Bacteriology.

We have had requests that these articles be reprinted in book form, thus giving a sufficiently complete work on the subject to meet ordinary requirements. Having secured a number of sets of the illustrative plates, we have decided to meet these requests. We trust our effort to give the medical profession a small but first-class work—both as to authorship, illustrations, press work, etc.—at a very reasonable price will meet with a reception such as we believe it to be entitled to.

The value of the original work has been much added to by the publication of an introductory chapter on "Bacteriology; its Practical Value to the General Practitioner," by W. Wayne Babcock, M.D., of Philadelphia.

BACTERIOLOGY:

Its Practical Value to the General Practitioner.

Bacteriology: Its Practical Value to the General Practitioner.

BY W. WAYNE BABCOCK, M.D.,

Pathologist to the Kensington Hospital for Women; Clinical Pathologist to the Medico-Chirurgical Hospital; Demonstrator of Pathology and Bacteriology in the Medico-Chirurgical College, of Philadelphia.

To the general practitioner, bacteriology is offering a constantly increasing field of usefulness. From its earliest days, this science has suggested a theoretical basis for treatment, while it has developed, especially during more recent years, numerous products of practical remedial value. For the most part it has not been difficult for the physician to avail himself of these advances in treatment. The methods founded upon theory have been particularly popular, and it has been only necessary for the micro-organismal nature of an affection to gain credence in order to have innumerable preparations of real or fancied antiseptic value pressed into service.

From the earlier vague and theoretic means of combating bacterial invasion, we are now emerging upon a more rational therapeusis, founded upon laboratory investigation and endorsed by clinical trial. Although preventive inoculations and the use of toxins and antitoxins have been adopted with some reluctance, their application is not difficult and their employment is now becoming general. Coincident with the recent advances in etiology, prophylaxis and therapeutics, and scarcely less important, are the strides made in bacteriologic diagnosis and prognosis. And yet, despite the great value of these latter innovations, it is entirely

7

probable that bacteriology has been of far greater service to the practitioner in the line of treatment than in diagnosis. The scratch of the vaccine lancet or the thrust of the antitoxin needle requires neither erudition nor great technical skill, while indulgent manufacturers beg to supply the practitioner with more convenient and refined products. On the other hand, unfortunately, the methods of diagnosis have not only required laboratory training but also laboratory apparatus.

These difficulties are being largely overcome in the medical centers by the establishment of municipal laboratories, but in the more remote districts the general practitioner is left, as usual, self-dependent. There is, moreover, a prevalent impression that the general practitioner requires but little knowledge of bacteriology. This needs correction. The specialist is a man who, by exceptional proficiency in a single branch, is exempted from a thorough knowledge of general medicine. Serving in a single field, he exacts service from his confrères in all others, including that of skilled laboratory workers. The true position of the general practitioner is, naturally, quite the reverse, and with the knowledge that "diagnosis is treatment," bacteriology must be far from the least of his many accomplishments. He may therefore rejoice in the fact that some of the most important of bacteriologic methods now require but simple apparatus, are capable of rapid performance and demand no exceptional skill. Given a good microscope, which may now be considered an essential part of the practitioner's outfit, a very moderate additional expense will provide the essential equipment for many of the very important diagnostic tests. Indeed, it is now possible without any aid from the microscope and with no more complicated apparatus than a test tube, to determine the presence and activities of typhoid bacilli in the body; and thus, by a procedure scarcely more difficult than the test for sugar in the urine, diagnose the existence of enteric fever. The single example may serve to illustrate the error of a prevalent opinion that bacteriology is hardly accessible to the practitioner as an aid in his daily work.

In this article I desire to mention a number of the practical advances in bacteriology and to indicate how much of essential value in the various fields they offer to the progressive physician. Certain of these advances may never have a general use; others are as yet very imperfectly developed; improvements may be expected upon all; yet we do our patients an injustice if we do not avail ourselves of some of the present benefits. It is convenient to group the bacteriologic advances under the general headings of *Etiology, Diagnosis, Prognosis* and *Treatment*. In attempting this, no excuse is offered for mentioning much that is trite to many medical readers.

Etiology.—On referring to the causal relations of bacteria to morbid conditions, we find that there is a large number of diseases for which micro-organisms may reasonably be claimed as the inducing factors; a considerable number for which certain bacteria have been described as the etiologic agents, and—until within a very recent period—but a very moderate number of which definite bacteria have been satisfactorily established as casual. Under this latter class we now have such prominent bacteria as Koch's bacillus of tuberculosis, Eberth's bacillus of typhoid fever, Klebs-Löffler's bacillus of diphtheria, Koch's spirillum of cholera, Neisser's diplococcus (gonococcus) of gonorrhea,

Nicolaier's bacillus of tetanus, Obermeier's spirillum of relapsing fever, and Kitasato's bacillus of plague. Scarcely less important and also well authenticated is the bacillus anthracis of malignant pustule, bacillus mallei of glanders, bacillus leprae of leprosy and the streptothrix (?) (ray fungus) of actinomycosis, and of madura foot. To these should be added the well-recognized micro-organismal causes of many septic and suppurative conditions—such as the varieties of staphylococci, so frequent in abscesses; the streptococcus found in erysipelas; the bacillus pyocyaneus of green pus; the bacillus coli communis, frequent in abscesses adjacent to the intestines; the bacillus aerogenes capsulatus, found in a usually fatal form of sepsis associated with gaseous edema, etc.

Nearly all of these organisms have been proved to be the specific causes of their respective diseases by conforming to the well-known rules of Koch. That is, the organism is found to be constantly present in the bodies of those affected by the disease, may be grown in pure culture outside of the body, when properly inoculated in lower animals it reproduces the disease, and may again be recovered from the animal so infected.

There are also good reasons for accepting a number of other organisms as specific. Recent investigations corroborate the claims of Sanarelli that his bacillus icteroides is the cause of yellow fever.

The bacillus described by Canon and Pfieffer for influenza, and by Canon and Pielicke for measles, are generally accepted. The Lustgarten bacillus of syphilis has not been sustained, but van Niessen claims to have cultivated from syphilitic blood a bacillus producing characteristic lesions in pigs and monkeys.

The claims made that the bacteria causing mumps, scarlatina and whooping-cough have been isolated, do not, as yet, appear conclusive; while the organisms of such notably infectious diseases as typhus and small-pox await discovery. The better knowledge of the bacteria producing the infections has been fruitful in improving hygienic measures—a subject too extensive to be considered at this time. Not less important have been the reflections of etiologic advances in improvements of diagnosis, prophylaxis and treatment.

Bacteriologic Diagnosis.—Having ascertained that a micro-organism is invariably found in the body in association with a single disease and with that disease only, the diagnosis of the disease may readily hinge upon the determination of the presence of the bacterium. The chief methods used to determine the character of a micro-organism, and thus diagnose the disease, depend upon: (1) Peculiarities in form and arrangement; (2) peculiarities of staining; (3) peculiarities of cultural growth; (4) peculiarities of effects produced when introduced into the bodies of certain animals; (5) peculiarities in behavior of cultures of the micro-organism when brought in contact with certain fluids from the diseased animals.

Certain of the higher forms of the vegetable parasites, especially the parasitic moulds and yeasts, are readily recognized by their peculiarities of form alone. For example, the saccharomyces albicans of thrush and the varieties of aspergilli found in the external ear are readily determined by microscopically examining some of the suspected material in a drop of an indifferent fluid; while an accurate diagnosis of the varieties of ringworm, of favus, of tinea versicolor, may be obtained by removing some of the affected epithelial scales and hairs and examining under the microscope for their respective moulds, after treating with alcohol and ether, and strong caustic potash to remove the fat.

The similarity between different varieties of bacteria is so great, however, and pathogenic micro-organisms are so frequently aped in shape and grouping by the benign varieties, that this first method of diagnosis, as a rule, is merely corroborative to staining and cultural methods. Thus the most noteworthy peculiarity of the tubercle bacilli is their resistance to decolorizing agents; and yet it is also very important that the organisms appear as short rods, with rounded ends, a slight curve and a tendency to be arranged in II, V or X like figures. The simple method of diagnosing this bacillus by staining has proved of chief value in tuberculosis of the respiratory tract, the presence of tubercle bacilli in the sputum being practical proof of the existence of the disease. The same method of diagnosing tuberculosis in other parts of the body has given less satisfaction. In tuberculosis of the genito-urinary tract, it is often difficult to find the bacilli in stained preparations of urinary sediment, and the bacilli are rarely seen in the pus from tuberculous abscesses. Although sputum that has become putrid may continue to show the characteristic bacilli, it is advisable to make preparations of fresh fluids. Urine, especially, should be speedily centrifuged and examined, and it is probably an advantage to use a method which obviates certain of the extraneous urinary substances. While the smegma bacillus and the leprosy bacillus resemble the tubercle bacillus in staining properties, it is to be recalled that both of the former bacilli stain more rapidly, decolorize more easily and have different grouping tendencies than the latter. The failure to find tubercle bacilli, of course, does not necessarily negative the presence of the disease, nor may the number of bacilli found bear any relation to the extent of the disease process.

Gonorrhea, at present, is most accurately diagnosed by staining. The organism, which is a rather large diplococcus with flattened surfaces in opposition, and a tendency to be found in groups in or upon the cells, is differentiated from most other diplococci by failing to stain by Gram's method, although readily coloring with the ordinary basic anilin dyes. In subacute or chronic cases the gonococci, having invaded the deeper epithelial layers, may be absent from the discharge except during exacerbations. It has, therefore, been suggested that in gleets of obscure nature the "Beer Test" be used; the visit to Bacchus being frequently rewarded by the return of the gonococcus to the discharge.

The staining of cerebro-spinal fluid withdrawn by the lumbar puncture has been a decided aid in diagnosing certain varieties of cerebro-spinal inflammation. The diplococcus intracellularis meningitidis of Weichselbaum, frequent in the epidemic and very fatal form, resembles the gonococcus by its presence in the leucocytes and by being decolorized by Gram's method. The diplococcus of pneumonia has also been found in the fluid of many of the fatal cases.

In the diagnosis of gastric carcinoma probably the most suggestive addition to the stomach contents, apart from the presence

of detached bits of tumor, is the long immobile, club-shaped
Oppler-Boas bacillus.

The most important recent advance in the diagnosis of puerperal sepsis is the employment of bacteriologic methods to as-
certain the variety of the infecting organism. The fatal infections with the gas bacillus (bacillus aerogenes capsulatus of
Welch) and with the septicemia-producing streptococcus should
be carefully separated from the milder infections by the colon
bacillus and the staphylococcus and the usually localized
gonococcus invasions. A sterile glass tube may be inserted into
the uterine cavity, some of the lochia withdrawn into the tube,
and the tube ends sealed pending the examination of the contents.

In general surgery the value of staining the fluids from infected foci, even during the operation, in order to determine the
necessary operative procedure, has not diminished, although the
method is not new.

Certain bacteria exhibit, under varying conditions, altera-
tions in their staining properties. Such bacteria may be more
distinctively stained after cultivation upon a particular medium.
This is so important with the diphtheria bacillus that the diagnosis is made by staining after cultivation upon Löffler blood-
serum mixture. The rapid growth, at incubator temperature,
of the irregularly rounded, elevated, porcelain white colonies
tending to rapidly coalesce into a diffuse smeary layer, is also
quite characteristic of this organism. Elsner, Kashida and Hiss
have devised special media[1] for developing characteristic cultures

---

1 Piorkowsky (*Ber. klin. Woch.*, 7, 99,) has recently recommended a medium
composed of normal urine containing 0.5% peptone and 8.8% gelatin. The medium is
inoculated from the stools, and incubated 90 hours The typhoid colonies appear as
colorless spots of radiating threads, while the colon cultures are sharply defined,
round, yellowish colonies.

from the typhoid bacillus, but their practical diagnostic value can hardly compare with the simple method of serum diagnosis. Cultures, at present, are of chief value to the practitioner in diagnosing diphtheria, and those who have not the advantage of a laboratory may procure the necessary media, and with a little training make their own examinations. In lieu of an incubator the inoculated tube, in a protecting case, may be carried in an inner pocket, the body supplying the necessary warmth.

Animal inoculations have proven useful in diagnosing genitourinary tuberculosis in cases where the bacilli were not shown by staining. Some of the fresh urinary sediment (obtainable from a single kidney by the ureteral catheter) is injected into the abdominal wall of a guinea pig that is killed after five weeks and examined for tuberculous lesions. Animal inoculations are also of value in diagnosing the type of many of the infections; while the subdural inoculation, in a rabbit, of a bit of the spinal cord of the affected animal remains the best method of ascertaining the existence of rabies.

One of the greatest advances in bacteriologic diagnosis is that dependent upon the tendency of the bacteria of a given disease to agglutinate, or collect in clumps, and to lose their motility when brought in contact with the serum from a person affected by the disease. Although this method is useful in determining the variety of the bacteria, the presence of infective organisms in drinking water, etc., its greatest value has been in diagnosing disease.

Being present in over ninety-five per cent. of cases of typhoid, it is the most useful diagnostic sign yet discovered. The method has not been as thoroughly developed in other diseases, but has shown a value in aiding the identification of cases of cholera, glanders, malta fever, tuberculosis, leprosy, relapsing fever and other diseases. Considering that the method is yet in its infancy, the results obtained have been surprisingly accurate. The test is performed with or without the microscope, the former having the advantage of quickness and greater accuracy. The main difficulty that the practitioner will find with this method is the difficulty of keeping on hand fresh cultures of the bacilli. The advantages of emulsions of dead bacilli are readily apparent, but thus far they have proven less reliable than the fresh cultures.

A final, simple and practical method of diagnosis is by the injection of bacterial products into the affected animal. Tuberculin, a glycerin extract of tubercle bacilli, has proved to be a most reliable agent for diagnosing tuberculosis. Unfortunately, its tendency to exacerbate this affection has largely precluded its use in the human family. Mallein has also proven successful in diagnosing glanders in horses.

Prophylaxis and Treatment.—Two main classes of bacterial remedies have been developed, namely, those obtained directly from the micro-organism and those indirectly obtained by the action of bacterial products upon animals. To the first class belong the toxins and vaccines; to the latter class the antitoxic and bactericidal serums. The former have been of chief value as prophylactic and immunizing agents, but, as yet, have not revealed the curative powers of the latter class. Upon the other hand, the antitoxins have more transient immunizing powers than the toxins or vaccines. Protective inoculations or vaccinations are now accessible to the practitioner for small-pox, anthrax, cholera, plague, hydrophobia, and for a number of diseases of the lower animals—as black leg (symptomatic anthrax), hog cholera, etc.

The curative effect of the bacterial products of the tubercle bacillus devised by Koch have hardly proven commensurate with their dangers, while the employment of Coley's mixed toxins of prodigiosus and erysipelas for the relief of inoperable malignant tumors has been followed by a cure only in occasional cases. The antitoxins have shown temporary but valuable immunizing powers in diphtheria and tetanus, and the diphtheria antitoxin has revealed marked curative powers—much greater, indeed, than that of tetanus.

The use of streptococcus antitoxin has been followed by good results in certain cases of sepsis and even small-pox. The results suggest the desirability of greater accuracy in diagnosing the variety of the infecting organism as an aid to the proper treatment.

With a prospect for the future of increased and simplified diagnostic advances, and multiplied and more efficient curative products, the advantage of a more thorough knowledge of bacteriology to the practitioner are not easily overestimated.

AN ATLAS

OF THE

BACTERIA PATHOGENIC IN MAN.

An Atlas of the Bacteria Pathogenic in the Human Subject.

PART I.

BY SAMUEL G. SHATTOCK, F.R.C.S.

THE object of the accompanying plates is to present graphically from original preparations all the chief bacteria which are pathogenic in the human subject.

The originals of the drawings (in the Museum of the Royal College of Surgeons, London) have been made under the author's supervision by Mr. G. T. Gwilliam, F.R.A.S., who, from an astronomical experience, is a draughtsman of extreme accuracy, and they may be relied upon as absolutely faithful representations of the objects themselves.

All the preparations have been drawn as viewed under a $\frac{1}{12}$ homogeneous immersion, Leitz, ocular No. 2, tube length 170 millimeters, giving a magnification of 680. It is to be noted, however, that in order not to tax the eye, they have in all cases been slightly enlarged in the drawing. The actual magnification amounts to about 1,000.

Each illustration gives the chief forms selected from several fields in order to bring out their variations, their differences in size, or their grouping, etc., a result which will dispel the idea of uniformity in these respects that is sometimes gained from representations where such differences are studiously excluded.

All the drawings have been made from cultures in an active

21

stage of growth, in order that abnormal, degenerate forms might be eliminated. The precise mode of preparation, the age of the culture, and, when important, its original source, are given under the several figures.

The fungi, which are pathogenic in the human subject, belong to the botanical subdivision of (1) Schizomycetes (Σχίsω, I split, Κύκης, a fungus) Fission-fungi, or Bacteria; (2) Blastomycetes (Βλαστος, a bud) Budding-fungi, or yeasts; (3) Hyphomycetes ("Υφος, a web), Hyphal-fungi, or moulds.

Strictly speaking, the study of bacteriology does not comprise that of yeasts and moulds, which are for this reason excluded from some systematic works devoted to bacteriology. Much less could the consideration of animal micro-parasites find a place in a strict system of bacteriology, although, owing to its present small dimensions, this subject is included, and conveniently so, in at least one standard English work on "Bacteriology."

With the exception of thrush, all the illustrations given are limited to bacteria proper or schizomycetes.

In regard to this classification of pathogenic fungi,* it will be enough to say that bacteria are achlorophyllous vegetable organisms, characterized by a fissiparous method of reproduction, though in a certain number there occurs a second or additional method, viz., spore-formation.

Yeasts, whilst again they may multiply by spore-formation, do so as an ordinary method by budding, i.e., in place of the cell becoming symmetrically partitioned into two or more ele-

---

* The absence of chlorophyll, which has commonly served as a basis for defining the group of fungi, is not held by Sachs as sufficiently fundamental for such a subdivision; he classes with achlorophyllous schizomycetes certain chlorophyllous forms which multiply by a similar method, under the name of schizophyta, etc.

ments, there grow out one or more processes or buds, which, after increasing to a certain size, may become abstricted off and disconnected from the parent cell, or whilst acquiring an independent cavity, may remain connected with it and generate further elements by the same plan. The hyphomycetes occupy a higher position than either of the foregoing groups, in that they present distinct organs of fructification, and are capable, moreover, of a proper sexual process of propagation.

Not that these subdivisions are more clearly defined than are others in biology.

The branching filamentous forms, at times met with in the tubercle and tetanus bacillus, make an approach toward the mycelial productions of moulds; and in actinomyces and streptothrix Madurae, a dense and branched mycelium is regularly developed in artificial cultures.

The microscopic characters of bacteria are in scarcely any case sufficient to allow of their identification. It is by the *ensemble* of their cultural characters on different media, their chemical products, the results of experimental inoculation upon animals, that identification is possible, and to these must be added the effects produced upon bacteria by the action of specifically immunized sera.

Accumulated observations have disclosed a bacterial kingdom of such dimensions that to determine or identify any given individual is a matter of increasing difficulty. The problem of identification is reduced, however, within comparatively easy limits when those organisms which are pathogenic are alone considered, and still more so when attention is restricted to those that cause disease in the human subject.

The fundamental argument adopted by Darwin for the

theory of evolution, viz., the impossibility of defining species, holds equally through microscopic and macroscopic forms of life. Nevertheless, of the different methods of classifying bacteria, those founded upon form are the only ones at present practicable. Of such classifications there are almost as many as there are authors. For all practical purposes, however, the pathogenic bacteria are reducible to spheres and rods; the latter being of the most various lengths, at times produced into long filaments, which may be straight or spirally twisted, or, in a few instances, branched.

As to the terms bacterium and bacillus, the former is now obsolete; every straight, rod-shaped organism is a bacillus. If the first is still employed, it is only in deference to pre-established usage; the early distinction between the two, founded as it was upon difference in length, is rendered valueless by the variations in this respect presented by one and the same organism; in the more highly pleomorphic examples, the shortest forms of bacilli may be indistinguishable from spheres or cocci. The proposal to name as " bacilli " straight, rod-like organisms which form spores, and " bacteria " those that do not, has not been followed, since there is nothing whatever in the terms indicative of such a difference. Besides the straight rod or bacillus, the other chief forms of pathogenic bacteria are the spherical, or cocci; and the twisted rod or spirillum.

It may be observed, in passing, that micrococci are not in all cases geometrically spherical. The faces of subdivision are very frequently flattened so long as the resulting elements remain united, whatever disposition the grouping takes—whether in twos (diplococci), fours (tetracocci), eights (sarcinacocci), or linear series (streptococci); it is not rare again to find the elements of a chain—or streptococcus elongated in the direction of its length, *i.e.*, oval in place of spherical.

The *microscopical examination of bacterial cultures* may be reduced to a comparatively simple technique. Bacteria may be examined after being dried, or in their natural condition. The advantage of the first method is that the preparations are permanent, and to this must be added the greater facility with which they admit of being drawn or photographed. Under the other, the organisms suffer no reduction in size or alteration in form, nor can appearances artificially produced be mistaken for such as are natural. There is the same difference, in short, as between the study of plants growing in a garden and those dried in a herbarium. The second method is, therefore, in all cases preferable; obviously for the study of growth and reproduction it is absolutely necessary.

The Hanging Drop.—Examinations in the natural state are carried out by means of the hanging drop, which may be described in detail as being the most beautiful of all methods, and as expeditious as any.

If a culture on agar or gelatine (when this is not liquefied by the growth) is to be examined, a thin, narrow ring of vaseline is painted around, but a short way from the depression of a hollow ground slide; a single öse or loop of distilled water is then transferred to the centre of a square or circular cover-glass (No. 1 thickness). For the sake of those unacquainted with such matters, it may be said that the öse or loop is a small eyelet made at one end of a piece of platinum wire about two and a half inches in length, the other end of which is pressed into the end of a piece of solid glass rod, seven inches long, softened in the flame of a Bunsen burner; the diameter of the eyelet, which is made by bending the wire with the end of a small-pointed pair of forceps into a complete circle, should be two millimeters (about $\frac{1}{10}$ inch), and in all cases it is to be sterilized by raising it to a red heat in the flame before and after use (*Fig.* 1). After

Fig. 1.—The öse or platinum loop, the *length* of the wire and glass rod reduced.

sterilization in the flame, the loop is inserted into the culture tube, applied to the growing edge of the culture, withdrawn, and immediately transferred to the drop of water on the cover-glass, in which it is moved until slight turbidity appears; the excess of the culture on the loop is now burnt off in the flame, and the bacteria thoroughly distributed on the cover-glass by its means, the drop being extended so as to cover an area of about four millimeters in diameter. The cover glass is then seized at the edge in a small pair of forceps, turned over, and applied to the hollow ground slide so that the minute drop of fluid lies over the center of the hollow space in the slide without coming into contact with the latter; the gentlest pressure made around the border of the cover-glass will then render the moist chamber completely air-tight (*Fig.* 2). A moist chamber may be impro-

Fig. 2.—Section of hollow ground slide with cover-glass *in situ*, slightly reduced.

vised, however, in a very effective manner on an ordinary flat slide, as follows: Cut a piece of thick blotting-paper an inch square, fold it, and cut away a small semi-circle from the center of the folded border, open it out, wet it, and lay it on the slide;

the cover glass having been prepared exactly as before is inverted so that the hanging drop lies over the centre of the circular hole in the blotting paper (*Fig.* 3).

Fig. 3.—Section of slide and moist chamber prepared with blotting-paper slightly reduced.

If the examination be prolonged, it is necessary to moisten the paper from time to time. For the study of such preparations, much of the light must be cut off with the diaphragm, assuming that the Abbé condenser and plane side of the mirror are used; unstained organisms are otherwise scarcely discernible.

The most suitable area for observation is the edge of the drop where the film of fluid is thin, and where there is what may be called a "still" layer in which an accumulation of the bacteria takes place, and where their movements (whether active or Brownian) are less lively, and eventually cease.

The best plan of finding the edge is to place the slide with the hanging drop centrally on the stage of the microscope, transfer a droplet of cedar oil exactly over the hanging drop, and screw down the coarse adjustment until the oil rises up to meet the object glass ; the rest of the focussing is then carried out by means of the fine adjustment which is screwed until some of the bacteria come into focus ; the slide is then moved in any one definite direction until the edge of the drop is reached. Besides the forms, the motility of the bacteria is to be determined by the hanging drop. For the latter purpose, the use of broth cultures is preferable to those made on agar ; nevertheless, if a preparation of the typhoid bacillus, *e.g.*, is made from au agar culture of twenty-four hours' growth, in the way described, there is no difficulty in observing the motility of the microbe, though the movements are not so active as in a broth culture. Examined in the hanging drop, it is to be noted that even non-motile bacteria present marked passive or Brownian movements; these are distinguishable by the fact that no real loco-motion takes place, the movements being of a to-and-fro or oscillatory character.

Coloration of the Hanging Drop.—The hanging drop having been so viewed, may next be advantageously stained. It is not necessary for this purpose to make a new preparation. The cover-glass is removed and turned over on a piece of filter paper, so that the drop is again brought uppermost.

The best staining fluid is a dilute solution of gentian violet, made by adding .5 c.c. of an aqueous solution of gentian violet (gentian violet 2.25 grammes, distilled water 100 c.c.) to 10 cubic centimeters of distilled water. The dilution may be arrived at sufficiently well by placing a few drops of the gentian violet solution in a test-tube and adding distilled water until the fluid is sufficiently dilute to allow of being seen through when held between the eye and the light. The sterilized öse is dipped into this, and the drop of dye mixed on the cover-glass with the hanging drop already there; the cover-glass is then again inverted and replaced over the hollow slide or upon the blotting-paper of the moist chamber. The ordinary histological method of running in the dye from one side of a wet preparation made without hollow slide or moist chamber, is followed by too much displacement.

The action of the dye is easily studied in all its grades in

PLATE XXIII.

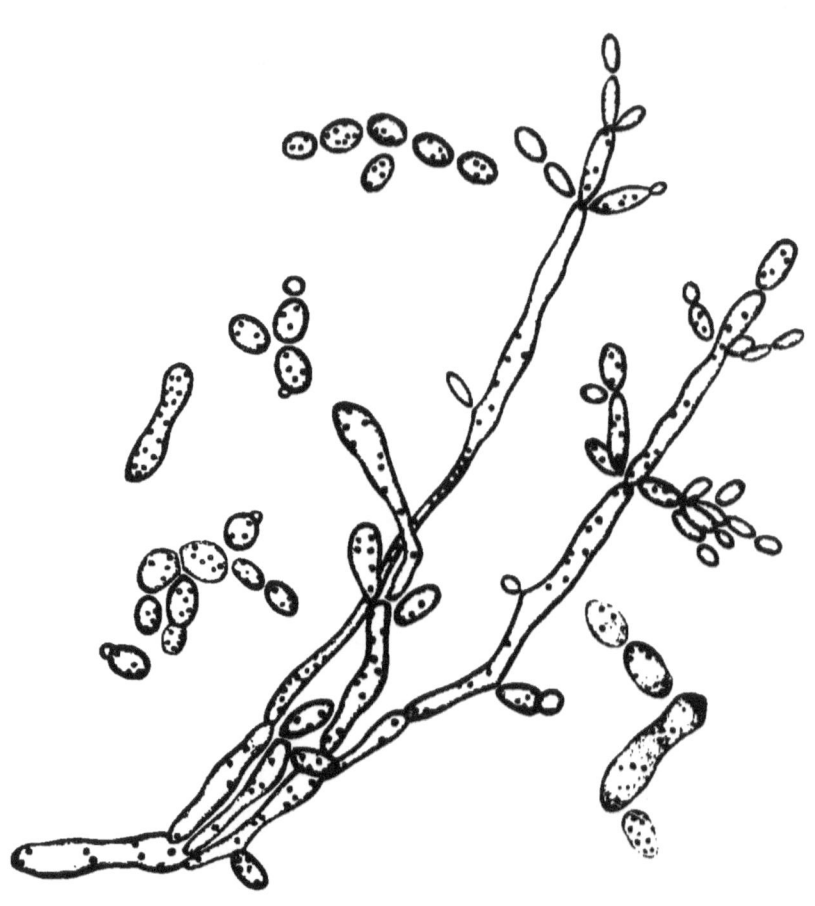

Saccharomyces Albicans.

PLATE XXIII.

SACCHAROMYCES ALBICANS. THE BLASTOMYCES OR YEAST-CAUSING THRUSH.

The drawing is made from a hanging drop of a broth culture which had been incubated at 37° C. for 24 hours.

There is shown on the left and upper part of the figure a series of simple oval cells; and below these, similar cells from which the formation of a second element is proceeding; this appears as a minute bud growing out from one of the ends or side of the parent cell. Where the cells lie in groups their opposed faces are flattened or facetted, an extensive group resembling a mosaic. The process whereby a filamentous stage is reached consists in the elongation of the cells, which continue to give rise to others from apex and sides until a complex branching system results. The chief secondary branches arise from the nodes of the main filaments, and from the former a tertiary set of smaller cells, and so forth.

On the right hand side are shown four cells, the protoplasm of which has been tinted with a dilute aqueous solution of gentian violet; they exhibit a certain number of spaces filled with fluid, or vacuoles, analogous to those holding sap in higher plants; and spherical granules, the function of which is not yet known,—possibly zymogenic in nature.

In the elongated or filamentous cells the vacuoles are sometimes of considerable length, and can be readily seen in the unstained condition.

PLATE XXIV

Fig. 1.—Typhoid Bacillus.

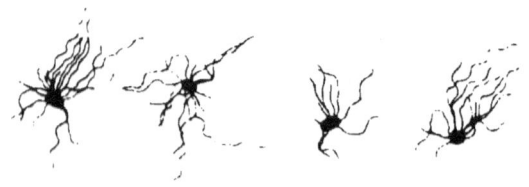

Fig. 2.—Typhoid Bacilli.

PLATE XXIV.

FIG 1.—TYPHOID BACILLUS.

From a culture 24 hours' old, incubated at 37° C., and made on the surface of agar.

Stained with carbol fuchsin, washed in water acidulated with acetic acid. Straight or very slightly curved rods, the shortest of which appear as oval: the bacilli present great variations in length and breadth: parallel grouping is fairly common. The filamentous forms exhibit no trace of a segmented or composite structure.

FIG. 2.—TYPHOID BACILLI.

Five typhoid bacilli, from a dried preparation stained by Van Ermengen's silver method, and showing the long wavy flagella arising around them. The largest number shown is seventeen, but this may be exceeded.

PLATE XXV.

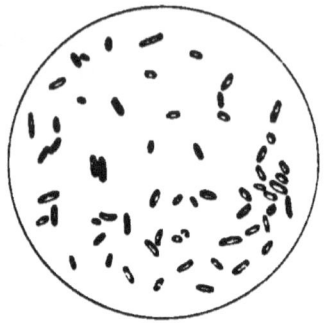

Fig. 1.—Colon Bacillus.

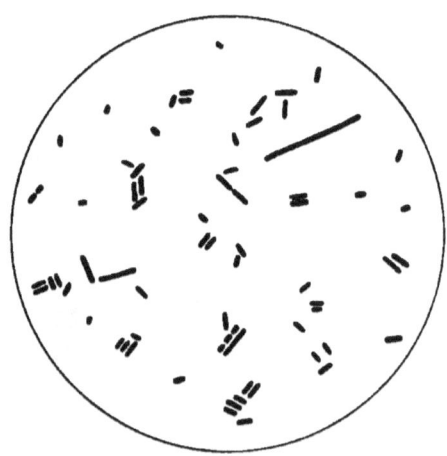

Fig. 2. Colon Bacillus.

MORRISON & GIBB LTD EDINBURGH

PLATE XXV.

FIG. 1.—COLON BACILLUS.

From a culture in a simple aqueous solution of Peptone, 24 hours' growth, incubated at 37° C.

The figure is introduced to show the "end-staining" often observed in bacilli, the causation of which is not fully understood. The interest attaching to the present specimen, is that it demonstrates the effect of one particular factor in leading to this result. It has been noticed by Mr. H. C. Haslam that the "end-staining" occurs regularly in the colon bacillus if it is grown in alkaline media.

In an unstained hanging drop corresponding appearances present themselves, the ends of the bacilli being darker and more granular than the centres.

FIG. 2.—COLON BACILLUS.

From an agar culture of 48 hours' growth, incubated at 37° C.

Showing the usual form of the Colon Bacillus: plump straight rods with rounded ends, variable in length and thickness, the shortest appearing as oval. Parallel grouping occurs. As there are many varieties of Colon Bacillus, it is necessary to take one as a standard; that selected is from a culture of Professor Escherich's (the original describer of the bacillus). A subculture of this particular organism was obtained from Professor Escherich by Mr. Wallis Stoddart, to whom the author is indebted for a sample of the strain.

PLATE XXVI

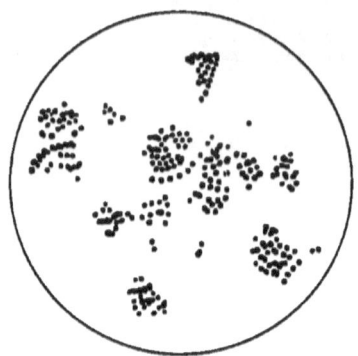

Fig. 1.—Staphylococcus Pyogenes Aureus.

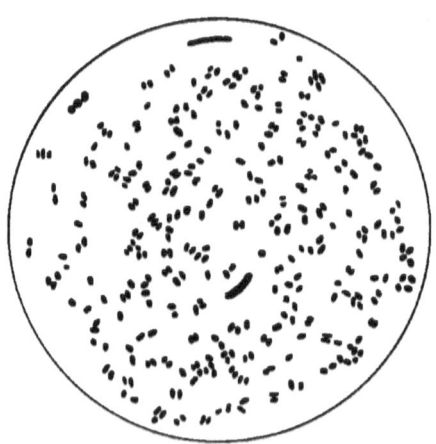

Fig. 2.—Plague Bacillus.

MORRISON & GIBB LTD EDINBURGH

PLATE XXVI.

Fig. 1.—STAPHYLOCOCCUS PYOGENES AUREUS.

From the edge of a growing culture on agar.

Carbol fuchsin, washed in acidulated water. The cocci occur in clusters or botryoidal groups (Σταφυλή – bunch of grapes), here and there in short chains, twos, or singly. The individuals in the groups vary in size, indicating probably differences in age; in some of the groups the process of division appears in a diplococcus with flattened sides, the elements of which have not yet parted and become spherical.

Fig. 2—PLAGUE BACILLUS.

Agar culture, incubated 37° C.; the advancing edge of a three days' growth.

Carbol fuchsin, washed in acidulated water. Short rods, mostly of oval form, often in pairs, side by side. Here and there slightly curved, longer forms occur, some of remarkable thickness.

The original source of the bacillus was a Lascar admitted into the Seamen's Hospital, Albert Dock, from a P. and O. steamer arriving from Bombay; the bacteriological diagnosis was made by Dr. R. T. Hewlett; inoculations carried out upon guinea pigs caused death with typical symptoms.

The forms examined in the blood of animals experimentally inoculated with this particular bacillus showed marked "end-staining"; this is entirely absent in the specimen prepared from the agar culture.

Plate XXVII.

Gonococcus.

A cover glass preparation of urethral pus, made by means of the loop, from the urethral meatus in a case two days after infection.

Carbol fuchsin, washed in acidulated water. Two flattened epithelial cells from the urethra, and one " polynuclear " leucocyte are represented. The stain has tinted the bodies of the epithelial cells. but not that of the pus cell. The cocci, without exception, occur in pairs with flattened or slightly concave faces; here and there is a group of four. The diplococci are set in groups or colonies which lie outside and unconnected with the cells; elsewhere, upon and within them. The groups connected with the epithelial cells are surrounded with an uncolored zone of ground substance indicating their location within the cell protoplasm. Those about the nucleus of the leucocyte are in cell protoplasm.

PLATE XXVII

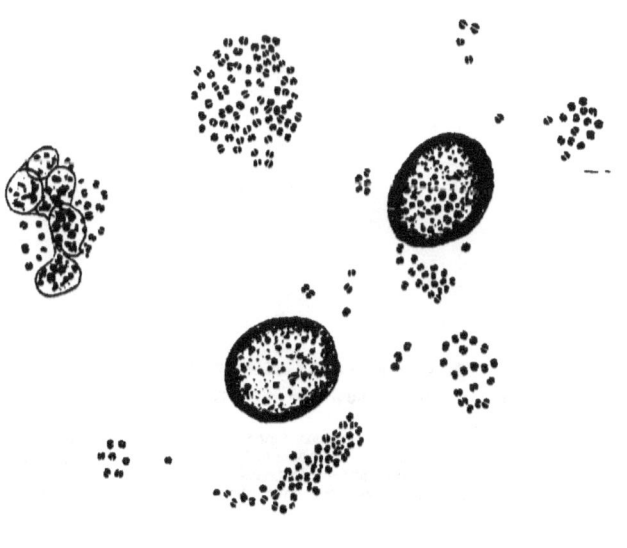

Gonococcus.

MORRISON & GIBB Lᵀᴰ EDINBURGH

Plate XXVIII.

Fig. 1.—Anthrax Bacillus.

Agar culture, incubated at 37° C.; of 24 hours' age : the films treated with 20 per cent. acetic acid before staining with aqueous solution of gentian violet.

The lowest filament is as yet sporeless ; some of its component segments are but half the length of others, and separated by a narrower interval, indicating that fission and interstitial elongation are proceeding in the filament.

In the others, various stages of spore-formation are shown. The spores appear at first as minute unstained points at or near the centre of the segments, the capsule of the spores preventing the penetration of the dye. When fully grown, the spores distend the wall or "sheath" of the bacillary cell, from which they are subsequently set free, as shown at the highest part of the figure. In many of the sporulating segments there is a minute second point, probably an abortive second spore, since the fully developed spore is in most instances not strictly central, but nearer one end of the spore-bearing segment, and this whether the latter presents a second point or not.

The staining of the spores is best carried out by means of hot carbol fuchsin, and the use of acid and methylene blue as described under the Tubercle bacillus.

Fig. 2.—Anthrax Bacillus.

Broth culture of 24 hours' growth, incubated at 37° C. Carbol fuchsin, washed in acidulated water.

Very few spores were found in the preparations, and none are present in the filaments figured. The spore formation takes place only under the free access of gaseous or atmospheric oxygen. The filaments adhere in tresses or strands.

Active subdivision is taking place in the segments, the ends of which remain straight or squarely cut.

PLATE XXVIII.

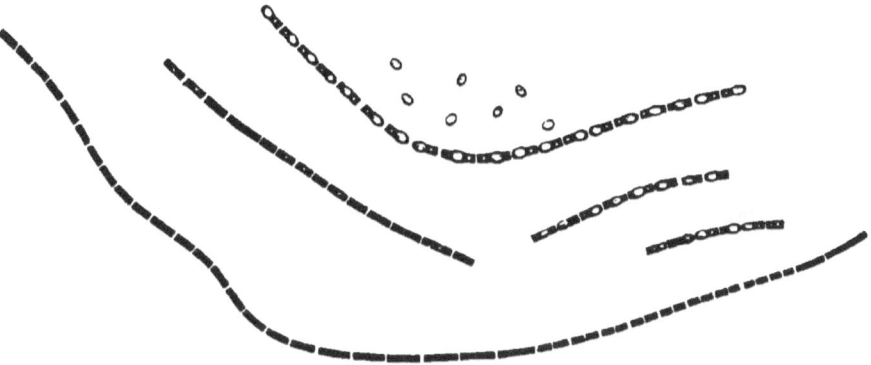

Fig. 1.—*Anthrax Bacillus.*

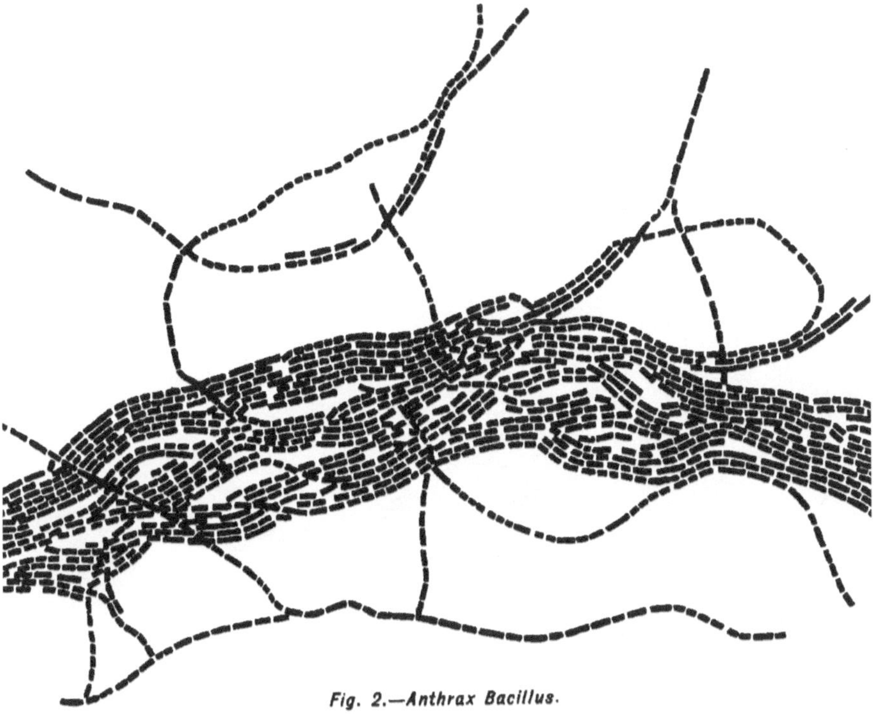

Fig. 2.—*Anthrax Bacillus.*

MEDICAL ANNUAL, 1898

MORRISON & GIBB L⁻ EDINBURGH

PLATE XXIX.

ANTHRAX BACILLUS.

"Impression preparation" of a small colony on agar in a Petri capsule; lowly magnified; the actual diameter of the colony is one millimetre.

Aqueous solution of gentian violet, washed in acidulated water.

The colony consists of a feltwork of wavy filaments, extending at the margins in tufts or hairy processes.

It was transferred to the cover-glass by allowing the latter to fall on the colony and then carefully raising it from one side.

PLATE XXIX.

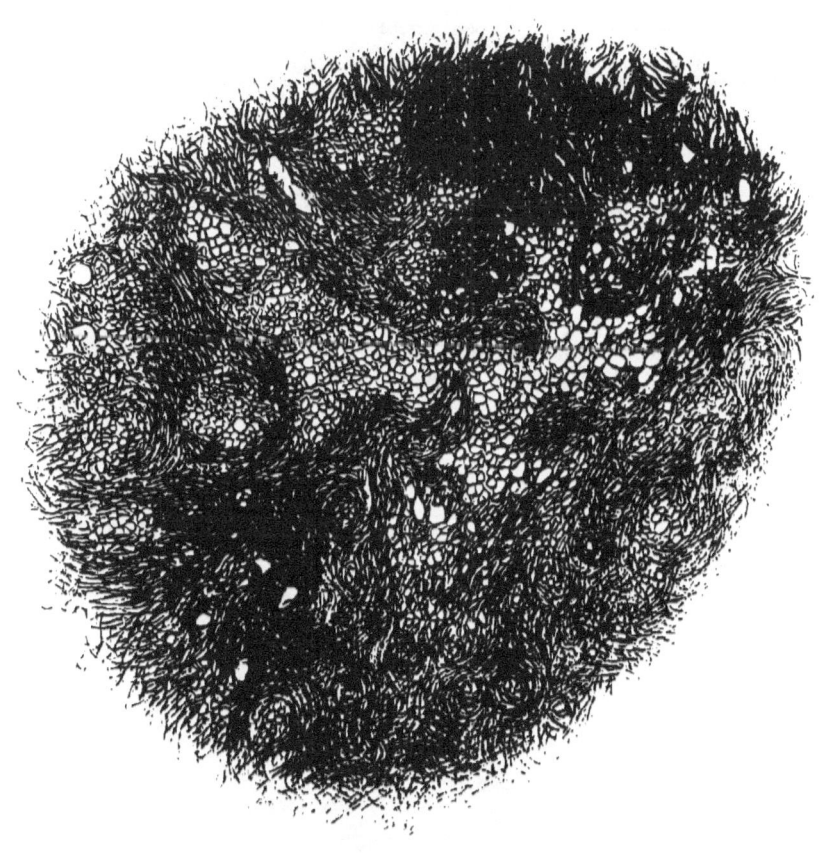

Anthrax Bacillus.

PLATE XXX.

TUBERCLE BACILLUS.

A preparation of Tubercular sputum, distributed over the cover-glass in the fresh state. Carbol fuchsin, heated on the cover, the latter then transferred to a watch glass of 25 per cent. hydric sulphate till decolourized, washed in water, and then counter-stained with aqueous solution of methylene blue in the cold, washed again in water, and allowed to dry. Tubercular sputum is best examined when no antiseptics have been used for its disinfection. It should be spread out in a thin layer over a sheet of glass on a black surface, or one of the black-bottomed plates made for the purpose; the most purulent foci are then isolated with needle and forceps. or cut out with scissors and forcibly broken up and distributed on a series of cover-glasses with the back of the forceps, whilst the covers are steadied by the pressure of a needle. Passing the films through the flame is not necessary. If the cover-glasses are not heated after the carbol fuchsin has been filtered on to them, the dye must be allowed to act for a considerably longer time—fifteen minutes: if heated, they must be held over a low flame in a small pair of forceps of which the ends have been somewhat sharply bent, in order that the dye may not be conducted to the under side of the glass.

The rationale of this particular method is as follows:—the carbol fuchsin stains all the different microbes present in such a film; with the exception of the tubercle bacilli, all are afterwards decolourized by the action of the hydric sulphate, to be re-dyed, together with the cell nuclei, by the methylene blue. This peculiar resistance to the action of the acid is common to the tubercle, leprosy, and smegma bacilli.

In the examination of pure cultures of the bacillus, this differential method is unnecessary, and the organism may be stained simply with carbol fuchsin, or by Gram's method, etc.

In addition to the "polynuclear" leucocytes (pus cells), there are shown in the figure two squamous epithelial cells derived from the upper part of the respiratory passages; the body of these is faintly tinted with the blue.

Many of the tubercle bacilli occur in small grotesque groups, that have been likened to Chinese characters.

The bacilli are mostly slightly curved, their extremities rounded, and their proto-plasm segmented, so that the microbes appear "beaded." The beading is not invariable in such preparations of sputum.

PLATE **XXX**.

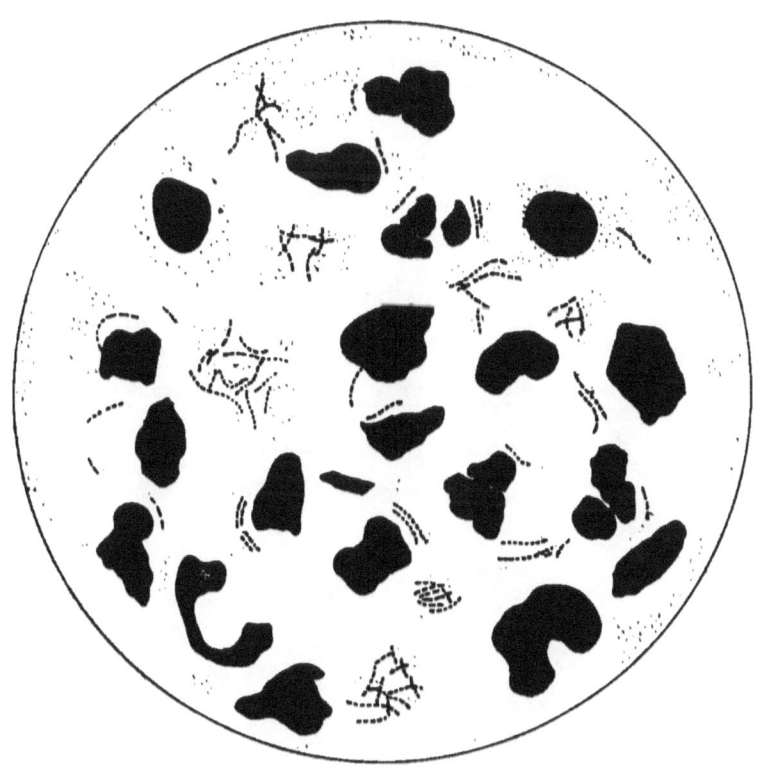

Tubercle Bacillus.

such a hanging drop, motile bacteria retaining their power of movement whilst the coloration is proceeding in them.

To Make a Permanent Preparation of the Hanging Drop.— Lastly, a permanent specimen may be made from the same preparation by removing the cover-glass, placing it again with the drop upwards on a piece of filter paper beneath a watch-glass or bell-jar, and allowing it to dry ; after which a small drop of Canada balsam (dissolved in xylol) is placed on a slide, a second small drop on the dried film, and the latter then allowed to fall on the slide in the usual manner. Or, after the examination of an unstained hanging drop, this may be allowed to dry in the same way and treated in the manner to be presently described under the dry method.

The dyeing of hanging drops made from broth cultures (in which case, of course, no water need be placed on the cover-glass) is not quite so satisfactory as those from agar, in consequence of an extremely fine colored precipitate appearing in the fluid, probably a proteid present in the broth, which unites with the dye. This precipitate is unconnected with the growth of the bacteria, as it appears if sterile broth is mixed with the diluted dye and examined as a hanging drop. Nevertheless, in such broth preparations when colored, many fields may be found devoid of adventitious granules, which, it may be noted in passing, exhibit lively Brownian movements.

The examination of jelly that is being liquefied by the growth of an organism is conducted precisely as that of a broth culture ; and admirable preparations may be obtained by coloring the hanging drop with dilute solution of gentian violet in the manner already described.

Of all the methods of bacterial microscopy, the hanging

drop, with few exceptions, is the best ; the important method of Gram, for instance, cannot be carried out except on a dry preparation.

The Dry Method.—The more minute attention paid to details, the more satisfactory will be the results.

To commence with, then, the cover-glasses (No. 1 thickness) must be properly clean. They are to be moved about in a capsule or other flat vessel of hydrochloric acid, washed in common water and then kept in a wide-mouthed stoppered bottle of absolute alcohol, to which glacial acetic acid has been added in the proportion of 10 per cent. The addition of acid will be found to prevent the cover-glasses receiving too high a polish when wiped, a result which greatly interferes with the equable distribution of the bacteria over the surface. Three cover-glasses having been wiped dry, a loop of distilled water is placed on each ; the loop should be drawn a short way over the surface, and if it is found that the water does not adhere to the glass, but returns again to a bead, the cover must be replaced in the acidulated alcohol and wiped afresh. The sterilized loop having been inserted into the culture tube and a small quantity of the growing edge withdrawn, it is applied to the drops of water in succession, and the residue having been burned off in the flame, the loop is used to distribute the turbid drop over the entire face of each of the cover-glasses, which are then allowed to dry; the distribution is facilitated by breathing on the surface, and the glasses are best steadied by the pressure of a needle. The prepared side is to be kept upwards throughout all the subsequent manipulations. The time-honored custom is now to pass the covers thrice across the flame, with the object of coagulating the albumen and fixing the films. Although so generally adopted, it is in reality wholly unnecessary; if the transit through the flame is made sufficiently slowly to really heat the glass, there is a risk of inducing still further shrinkage of the bacteria. Not only is the practice unnecessary for preparations made from agar cultures, but from careful comparative experiments, the author has found it equally without advantage in the case of such as are made from liquefied gelatine or from broth; the film will remain adherent even if the dye has to be heated on the glass, provided always that it is not afterwards violently washed beneath a stream, but in a vessel, of water; to such cultures the addition of water is not required; the loop is simply dipped into the fluid and distributed over the cover-glass.

The next procedure is that of staining. For this purpose aniline dyes are alone used; none others are sufficiently intense. The three chief are carbol fuchsin, aqueous solution of gentian violet, and aqueous solution of methylene blue; for special purposes modifications are necessary, the most important of which are Gram's method and those for staining spores, capsules, and flagella. Of these dyes carbol fuchsin* holds the first place, and takes the same position for bacterial staining that hematoxylin does for histological: it not only dyes all known bacteria, but gives, as a rule, cleaner preparations, in consequence of its being more readily washed out from extraneous material.

The cover-glass should be quite covered with a high bead of the stain. This may be done from a drop bottle, but the author invariably filters the stain directly on to the cover-glass through a small cone of filter-paper held across its overlapping parts in a pair of forceps. The stain having remained on from five to fif-

* Fuchsin, 1 gram: absolute alcohol, 10 c.c. Dissolve and add 100 c.c. of a 5 per cent. aqueous solution of carbolic acid.

teen minutes, the cover-glass is taken up with a fine pair of pointed forceps, tilted to throw off the stain, and at once gently submerged in a saucer of water faintly acidulated with acetic acid ; the washing may be completed in a second or a third saucer, and finally in unacidulated water. The degree of acidity can be arrived at by pouring a few drops of glacial acetic acid into a saucer, inverting it to get rid of the excess, and then filling with water, common or distilled—a solution of about .5 c.c. glacial acetic acid to 70 c.c. of water.

The washing is preferably carried out in this way rather than beneath a stream, as the latter tends to displace the film ; nor is the excess of the dye so well removed by water without the acid.

Fig. 4.—Sheet of lead bent at right angles, with filter-paper folded upon it, and cover-glass stood edgewise to drain without the film coming in contact with the paper.

After the final washing the cover-glass is allowed to drain and to dry by placing it obliquely with the prepared surface downwards, against a sheet of filter-paper folded at right angles over a piece of lead (*Fig.* 4). It may be more rapidly dried by pressing between filter-paper, but at the risk of damaging the film and transferring fibers to it. When dry, the preparation is completed by mounting it on a slide with a xylol solution of Canada balsam.

In a certain few cases (as in agar cultures of the anthrax

bacillus) this method does not satisfactorily clear the preparation, *i.e.*, decolorize the ground substance in which the bacteria lie.

Under such circumstances, the cover-glasses, after being prepared with the bacteria and allowed to dry, are to be left for ten minutes in a capsule or saucer of 20 per cent. solution of acetic acid ; after which they are washed in distilled water, allowed to drain and to dry, edgewise, on filter-paper, and afterwards stained in the usual manner, the subsequent washings being carried out without acetic acid. It is not *requisite* to pass the films across the flame at any stage of the procedure, even if the cover-glasses have been prepared from broth cultures. Even this at times fails to give clean preparations (as, *e.g.*, in the case of the glanders bacillus); the use of saturated alcoholic solution of gentian violet, left on the cover-glass for an hour (beneath a watch glass), will often then give satisfactory results ; the washing in such circumstances should be carried out rapidly in water to which no acetic acid has been added.

An Atlas of the Bacteria Pathogenic in the Human Subject.

PART II.

As stated in Part I., the object of this Atlas is to present graphically from original preparations the chief bacteria which are pathogenic in the human subject   The figures now intro-duced comprise nearly all the remaining bacteria of interest. Certain are omitted   Amongst such omissions may be named the bacillus of influenza, diplococcus intra-cellularis, bacillus enteritidis, bacillus melitensis, streptothrix pseudo-tuberculosa, etc.; and time will surely add yet others as causative agents, es-pecially in the group of infective granulomata.

Amongst the micro-organisms depicted in the present series, two of the most important in practical medicine are the typhoid* and diphtheria bacilli; and a succinct statement may be made in regard to the bacteriological diagnosis, firstly of typhoid fever, and secondly of diphtheria.

Typhoid Fever.—Although the bacteriological diagnosis of typhoid during life can be made from the discovery of the typhoid bacillus in the stools or the urine, such methods are of little practical avail, necessitating, as they do, a well-equipped laboratory and the expenditure of much time.

Widal's Reaction.—However interesting it might be to give an account of the observations of previous investigators which led up to Widal's application, neither space nor the main object of the present article will allow of it.   Widal was, however, the first to demonstrate that a reaction, which had been previously

---

* Other illustrations of this are given in Part I.

shown to take place between the serum of animals experimentally immunized against certain diseases and cultures of the specific bacilli producing such, could be obtained during the period of infection, and in this way serve as a means of clinical diagnosis. If the blood serum of an animal which has been rendered immune against, e. g., the typhoid bacillus be added in a test tube to a living broth culture of the same microbe, the bacilli in the culture rapidly cohere and subside to the bottom of the tube. This is known as "clumping" or "agglutination," and is taken to indicate the presence of an agglutinating substance or "agglutinin" which has formed in the blood in consequence of the presence of the bacilli experimentally introduced into the animal. The blood serum acquires this property, however, before immunization is established, i. e., during the progress of infection itself, and the property may be utilized as a means of diagnosis, for the important reason that the serum of a typhoid patient, whilst it will agglutinate typhoid bacilli, will not agglutinate those of other kinds. What is true in the case of typhoid is true also of other bacterial diseases. The phenomenon admits of two applications; a disease may be diagnosed by the action of the blood serum upon a known bacillus; or a bacillus may be diagnosed by the action of a known "immunized serum." It is to the former that importance attaches in practical medicine and surgery.

After this brief explanation of the *rationale* of Widal's reaction, the method of carrying it out may be described.

The broth culture of typhoid bacillus used must not be more than of twenty-four hours' growth, and must be grown in the incubator at 37° C, preferably in a broth that is not alkaline, but amphoteric (giving simultaneously an acid and an alkaline reaction to blue and red litmus paper). There arises even at this point the first possible source of fallacy. If such a broth culture be examined by the hanging-drop method (fully de-scribed in the previous volume), clumps of bacilli are not rarely encountered, and these, at times, of considerable size. In all cases, then, the broth culture is to be first passed through a small double cone of filter-paper into a watch-glass, in order to remove any such clumps present; and after this, a hanging drop is to be prepared for comparison with the result obtained when the reaction is tried for. One important difference to be ob-served in the microscopic preparations is that although in the broth culture clumps may be met with, the field between con-tains varying numbers of bacilli in active movement; whereas when clumping occurs from the action of the serum of a typhoid patient, the bacilli become motionless between the bacterial isl-ands, if any remain incoherent.

How to test the Blood Serum.—The reaction is obtainable quite early in the course of the disease, as early as the fifth day, and it persists during convalescence, but for an extremely vari-able period afterward. Not rarely the reaction will continue for a year, but it may last many years, and might without enquiry into a patient's history be erroneously taken to prove the exist-ence of typhoid on the occasion of some subsequent illness sug-gestive of it.

The blood to be used in the test is withdrawn either from the lobe of the ear or from the back of the finger near the root of the nail; and the puncture is best made by means of a surgi-cal or other neede with a cutting edge as well as a sharp point.

If the test cannot be carried out at the time, the blood must be collected in a pipette of the kind represented in the adjoining wood cut (*Fig.* 5), by breaking off the sealed ends of the

Fig. 5.—Pipette containining blood which has separated into clot and serum, the former occupying the lower half of the bulb (nat. size).

capillary tube on either side of the bulb, and applying one end to the issuing blood. When the expanded part is filled, the ends are hermetically sealed in the edge of the flame of a spirit-lamp. Blood so stored can be tested at leisure, and (if kept in the dark) retains its qualities for long periods. Even if dried the blood will provide the reaction; for this purpose it is collected on a series of cover-glasses, which, after being allowed to dry, may be, if necessary, posted to an expert for examination. If dried blood is used, a solution of the specific substances in it is obtained by means of distilled water; but the method is an inferior one, owing to the difficulty of estimating the dilution reached— a matter of cardinal importance.

Nine loop-fulls* of the filtered broth culture of typhoid are placed *separately* and fairly close together on an ordinary microscopic slide, the loop or öse being introduced as many times into the broth. The platinum wire is now sterilized in the flame, and with it a single loop of the blood serum is transferred to the slide and well mixed up with the whole of the nine droplets of broth culture. If the test is carried out on the spot, a few drops of blood may be allowed to flow from the lobe of the ear or finger, and to clot in a small test tube or a watch-glass; the serum so furnished will be ample.

---

* For a figure of the " loop," see p. 26.

If the blood to be used has been stored in a pipette, the two ends of this are broken off, and the contents blown gently on to a slide. A hollow-ground slide having been prepared with a ring of vaseline, and a clean cover-glass (before commencing the proceedings just described), a single öse of the admixed broth and serum is placed on the center of the cover-glass and gently spread out so as to cover an area about 4 millimeters in diameter; the cover is then inverted, placed over the hollow of the slide, and gently pressed at the margin so as to render the enclosed space quite air tight. The preparation is now placed beneath the microscope and and examined, with a $\frac{1}{12}$ homogeneous immersion, or a $\frac{1}{4}$ objective, which answers perfectly well for a study of the result. As the preparation is unstained, much of the light (if $\frac{1}{12}$ is used) must be cut off by means of the diaphragm—the bacilli are otherwise scarcely visible.

If Widal's reaction ensue, it is seen in the movements of the microbes becoming sluggish and ultimately ceasing, whilst they become at the same time aggregated into clumps of the kind represented in *Fig*. 1, *Plate XVIII.*

The time allowed for the observation should be half an hour. If no reaction has ensued within this time, the result is to be reckoned negative, and the existence of typhoid may be excluded, not with absolute certainty, but with very high probability. In the case of such a negative result, similar examinations must be repeated during the course of ·the disease, as the reaction, for causes not known, is in some cases delayed. If, however, the bacilli become motionless, even without any marked clumping, or if they become motionless, and clump, the result is to be reckoned as positive.

Before deducing the existence of typhoid in these circumstances, nevertheless, the result of further dilutions must be tested, for the reason that the reaction ensues in certain cases in which typhoid does not enter into the question, and, moreover, that the blood of healthy persons possesses at times an unusual degree of the agglutinizing power which is normally present.

The diagnosis, it may be said, becomes strengthened in proportion as the reaction persists on dilution. In certain instances it has been ascertained that a dilution of 1 to 5,000 will yet suffice for its production, and even a considerably further dilution than this. The test is to be repeated, therefore, with a dilution of 1 in 20 (*i.e.*, 1 part of serum to 19 of the broth culture), and if still observed, with a dilution of 1 in 50. The last limit may be held to suffice for the exclusion of other possible sources of phenomenon, and to establish a diagnosis of typhoid. The dilution of 1 in 20 is best made by placing nineteen loops of the broth separately on a slide, and mixing with a single loop of the serum; that of 1 in 50, by diluting 1 loop of serum with 10 of distilled water, and mixing one loop of this with four loops of the broth culture.

After use, all the slides, cover-glasses, and other materials are to be disinfected in a 1 in 20 carbolic acid solution.

Diphtheria.—For the bacteriological investigation of a supposed case of diphtheria it is necessary, firstly, to make a microscopic examination of a culture from the throat or nasal passages ; and secondly, if the investigation is to be complete, to inoculate animals with a pure culture in order to test the degree of virulence which the bacillus possesses, *i.e.*, both the morphological and the physiological characters of the microbe should be determined.

The inoculation of test tubes for the purpose of diagnosis

PLATE XVIII

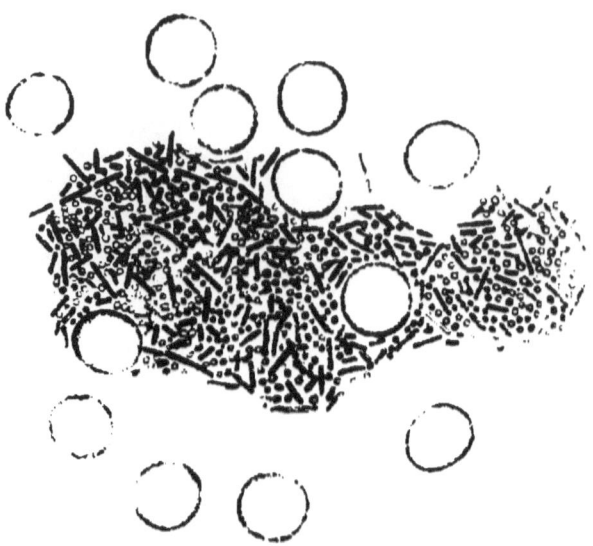

Fig. 1.—Widal's Reaction.

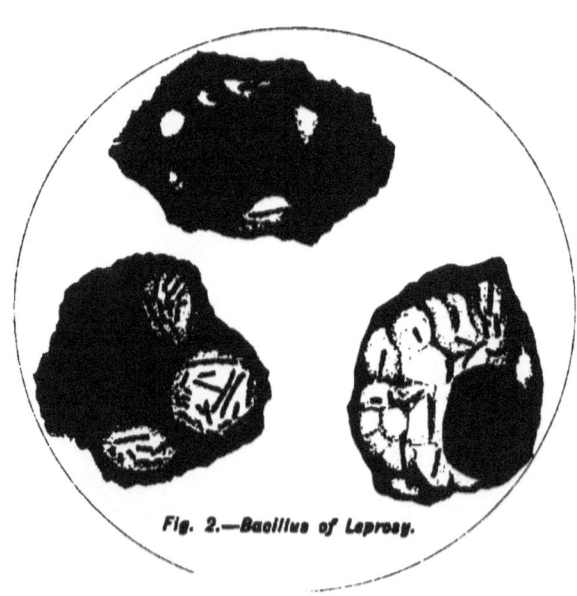

Fig. 2.—Bacillus of Leprosy.

SCOTT & FERGUSON, EDIN^r

PLATE XVIII.

Fig. 1.—WIDAL'S REACTION.

This figure is given to show the bacillary clumps which form when the blood serum of a patient suffering from typhoid fever is added to a living broth culture of the typhoid bacillus, the phenomenon being known as Widal's reaction.

In the case of the preparation figured, 1 part of blood serum was added to 19 of a twenty-four hours' (incubated) broth culture. The blood had been collected in a pipette and allowed to clot; a certain number of red corpuscles are admixed with the serum.

The clumping or agglutination of the bacillus is readily observable under 1-6th objective, though the clump represented was drawn under 1-12th oil immersion, and is magnified about 1000 times.

The method of carrying out the test for the diagnosis of typhoid fever is fully given in the text.

Fig. 2.—BACILLUS OF LEPROSY.

The figure represents three highly vacuolated endothelial cells from a lymphatic gland secondarily infected in a case of leprosy of the tongue. The cells occurred along with others of similar character in irregular groups scattered throughout the gland. The notable vacuolation (regularly seen in such "leprous cells") is possibly due to an abundant formation of digestive fluid secreted by the cell that it may destroy and utilise the bacilli. Highly vacuolated cells at times hold extremely few bacilli, possibly as a result of such a process of destruction. The bacilli mostly lie in the septa between or around the vacuoles, though when the vacuole is not viewed in strict optical section they appear to lie within.

The leprosy bacilli are slender, straight, or slightly curved rods, very uniform in breadth, and fairly so in length, and they closely correspond in size and general character with those of tuberculosis, as the latter are met with, e.g., in phthisical sputum. They present a markedly beaded appearance arising from protoplasmic segmentation. They give, again, the same common staining reaction, in this resembling, moreover, the bacillus of smegma; i.e., after being dyed with carbol fuchsine the bacilli resist decolorisation in a 25 per cent. mixture of sulphuric acid in distilled water. The smegma bacillus is not infrequently present in the urine of both sexes, and may be mistaken for that of tubercle. This error can be avoided by examining the urine drawn off by catheter, the smegma bacillus being in this way excluded. And to select one of many differential staining methods, though dyed with carbol fuchsine, the colour of the smegma bacillus is discharged in a mixture of 20 per cent. nitric acid in alcohol, whilst that of the tubercle bacillus is retained. For the reliable examination of urine a centrifuge is indispensable.

The sections of the leprous gland were stained for fifteen minutes (without heat) in carbol fuchsine, passed through 25 per cent. sulphuric acid, washed in water and counterstained for five seconds in a 1 per cent. aqueous solution of methyl blue, after which they were passed through water, absolute alcohol, oil of cloves, and mounted, finally, in a solution of Canada balsam in xylol. For the counterstaining of tissue *methyl* not methylene blue, is to be used; the latter is almost entirely removed by the subsequent immersion of the section in alcohol. The counterstaining of cover-glass films of phthisical sputum is satisfactorily carried out by means of an aqueous solution of methylene blue. (See the previous volume.)

The Leprosy bacilli are not in all cases located within the cells of a tissue: they may lie in a ground substance or glœa occupying the lymph spaces.

PLATE XIX

Fig. 1.—Diphtheria Bacillus (typical form).

Fig. 2.—Diphtheria Bacillus (atypical).

Fig. 3.—Diphtheria Bacillus (atypical).

Fig. 4.—Hoffmann's Pseudo-Diphtheria Bacillus.

SCOTT & FERGUSON, EDIN

PLATE XIX.

BACILLUS OF DIPHTHERIA.

Fig. 1.—THE "TYPICAL,' "KLEBS-LÖFFLER," OR "LONG" DIPHTHERIA BACILLUS.

From a culture on Löffler's blood-serum of twenty-two hours' age, incubated at 37° C. The growth was the third sub-culture of the original from the throat. Stained with Löffler's methylene blue, washed in tap water. The case from which the culture was raised concerned a boy (P. R.), admitted to St. Thomas's Hospital, October 18th, 1897, with inflamed throat; there was much membrane on the tonsils, and in the larynx as evidenced by the stridor and retraction of the chest. Tracheotomy was performed, the tube being removed on the fourth day. Four thousand units of diphtheria antitoxin were administered by subcutaneous injection. Until the early part of November progress had been favourable; on November 24th palatal paralysis was noted, the voice acquiring a nasal twang; this slowly improved. The knee-jerks were absent on November 29th, and still absent on December 24th, but the patient could at that date swallow without regurgitation.

The Bacilli are straight or slightly curved rods of varying length and thickness, often set in parallel groups of two or more. The designation of "long" implies that many, not all, are of conspicuous length. One end (or both) may be enlarged or bulbous; in the absence of this, the ends are abruptly rounded without tapering. The bacilli present a notable segmentation of protoplasm, which is divided into deeply stained parts; these, which may be far from equidistant, are in some instances flattened across the long axis of the rod, in others, spherical. The extremities of the bacillus correspond with terminal segments. In some of the shorter rods only terminal parts are dyed—end or polar staining.

The primary forms appear as short uniformly stained bacilli, with ends slightly smaller than the rest of the rod; these undergo transverse division, before the completion of which they appear as diplo-bacilli with the opposed ends flattened. Segmentation rapidly takes place, the shortest forms exhibiting only end-staining; the latter are distinguishable from true diplo-bacilli by the absence of tapering free ends and the length of unstained centre. These forms of the diphtheria bacillus are classed as "typical," since they are commonly associated with the typical, more virulent examples of diphtheria; yet not invariably so, for not only may "atypical" forms be highly virulent, but similar "typical" or "long" forms may possess little pathogenicity (as tested upon the guinea-pig), and occur in cases which clinically present no other features than those of sore throat, unaccompanied even with any marked malaise.

Fig. 2.—"ATYPICAL" OR "SHORT" DIPHTHERIA BACILLI.

A culture of thirteen hours' growth on Löffler's blood serum, carried on from one (Viennese) of the virulent strains in use at the Conjoint Laboratories of the Royal College of Physicians and Surgeons, London, for the preparation of toxin employed in the production of diphtheria antitoxin from the horse. Besides a few segmented forms are others of an "atypical" kind, commonly ranged in parallel collections of two or more; these mostly taper off at the extremities, presenting a deeply stained centre, which is usually divided by an uncoloured narrow line indicative of an incompleted transverse fission, the elements being double, or diplo-bacilli. Beyond the deeply stained centre the bacillus is of a lighter blue. Here and there a pyriform element occurs, more deeply stained towards the larger end; or short deep'y stained spindles.

Fig. 3.—A SECOND EXAMPLE OF THE "ATYPICAL" OR "SHORT" DIPHTHERIA BACILLUS.

From a virulent case of the disease. Diplo-bacilli in parallel groups; beyond their opposed deeply stained central ends the component elements taper off and are but lightly coloured. Pear-shaped forms are present, and a few which exhibit polar and segmented staining. From its staining reaction the form has been named the "sheath" variety by Dr. J. Eyre, to whom the author is indebted for the preparation from which the figure has been drawn. The "sheath" variety occurs, as a rule, in the milder types of diphtheria. It is but rarely met with, and is not strictly stable; if grown for some days upon alkaline potato and again sown on blood-serum, it acquires the segmented characters of the "typical" form.

Another atypical pathogenic variant has been described of the same form as the above, but staining uniformly and deeply throughout.

Fig. 4.—HOFMANN'S BACILLUS.

Culture of twenty-four hours' on agar, incubated at 37° C. Stained with Löffler's blue, washed in tap water. The culture was carried on from one isolated by Dr. E. A. Peters, to whom the author is indebted for a sample of the strain. Diplo-bacilli, occurring in parallel groups of two or more. The elements composing a single diplo-bacillus are short, squat, wedge-shaped, with opposed bases, and stain uniformly throughout. They are markedly shorter than the atypical or short varieties of the diphtheria bacillus, and relatively broader at their bases. In older cultures segmented and irregular involution forms may be encountered. Hofmann's bacillus (sometimes named a pseudo-diphtheria bacillus) is not found as a cause of true diphtheria in the human subject. Hence, though isolated in many forms of sore throat, such lesions are not to be regarded as diphtheritic. The bacillus, may, however, be associated with the "typical," "Klebs-Löffler," or "long" diphtheria bacilli in diphtheritic affections, but under such circumstances its presence may be regarded as of secondary significance.

PLATE XX

Fig. 1.—*Proteus Vulgaris*, ,

Fig. 2.—*Proteus Vulgaris (the early
filamentous phase).*

STOTT & FERGUSON, EDINR
[ILLUSTRATION CO. LTD]

PLATE XX.

BACILLUS PROTEUS VULGARIS.

This organism is introduced as one of the most common of those causing putrefaction, though different varieties of the bacillus coli are almost as ubiquitous.

Although by some, putrefactive organisms are not classed as pathogenic, because they are not the causes of any process differentiated clinically as a disease, they are pathogenic in the wider and truer sense, since putrefaction plays so important a part in the sepsis of wounds and in the toxæmia accompanying cancrum oris, the ulceration of extensive carcinomata of the alimentary and respiratory passages and conditions of a like kind.

Fig 2.

An " Impression preparation" from a gelatin culture of eighteen hours.

Carbol fuchsine, washed in water weakly acidulated with acetic acid. The culture was made by streaking a Petri capsule (after the jelly poured into it had set) with a straight platinum wire infected from a pure culture of the organism. The impression preparation is obtained by allowing a cover-glass to fall upon some part of the streak of growth and then gently raising it by one edge, when the line of culture is brought away adhering to the under side of the glass ; the specimen is then stained in the usual way.

The preparation shows the early or filamentous stage in the growth of the bacillus, which is highly pleomorphic, whence its name of Proteus.

The streak (of which one edge is depicted) consists of long, closely-matted, unbranched filaments arranged in strands. As seen at the free margin, some are sharply re-curved upon themselves.

The organism figured was isolated from macerating muscle, the actual material being beef steak, which was minced and incubated in distilled water at 37° C.

Fig. 1.

An impression preparation of a similar streak culture at a later stage when the gelatin was in process of liquefaction. Carbol fuchsine, washed in acidulated water.

The filaments have mostly divided into short rods, often constricted across the middle or in pairs end-to-end as diplo-bacilli. Here and there longer rods occur and filamentous forms, but the longer of the latter have not been introduced.

PLATE XXI

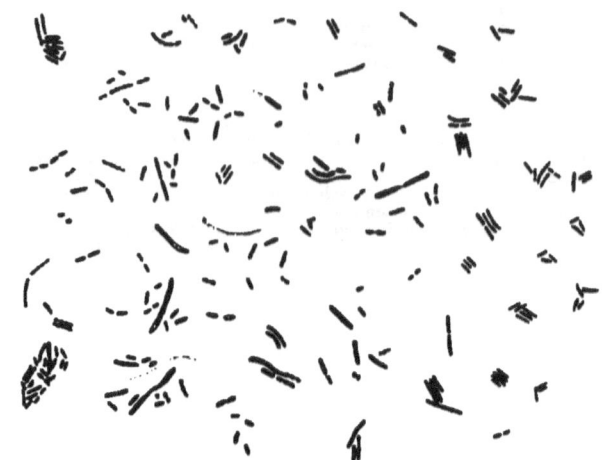

Fig. 1.—*Bacillus of Glanders.*

Fig. 2.—*Bacillus Tuberculosis, in a pure culture.*

SCOTT & FERGUSON, PRINT
EDINBURGH & CO. LTD

Plate XXI.

Fig. 1.—BACILLUS OF GLANDERS (BACILLUS MALLEI).

From the growing edge of a culture on glycerin-agar of three days' growth, incubated at 37°C.

Carbol fuchsine, washed in diluted acetic acid. The culture was made from a potato growth of characteristic honey-yellow colour, raised from a glandered horse, the potato culture being the second remove only from the original.

The bacillus (which is very difficult to stain with certainty in the tissues) may be stained in cover-glass films with aqueous solution of gentian violet (ten minutes), followed by washing in 1 in 10,000 caustic potash solution and afterwards tap water ; or, by means of alcoholic solution of gentian violet for one hour, rapidly washed in water. The action of Löffler's blue, for cover-glass films is uncertain.

Somewhat slender rods of varying length and thickness, straight or slightly curved, the shortest appearing as ovals, the longer as unsegmented filaments which are commonly less deeply stained ; some of the filaments present were longer than those figured. Here and there end-to-end pairs are met with. In some of the rods one or more minute, sharply-defined, circular vacuoles are present. A few of the bacilli present a moniliform outline, but none any distinct segmentation of protoplasm or "beading," as they may in the tissues.

The organisms tend to cohere in clusters, some of the smaller of which are selected in the illustration.

Fig. 2.—TUBERCLE BACILLUS.

Pure culture on Löffler's blood-serum, incubated at 37°C., carried on from a growth on glycerin-agar which was raised from the lymphatic gland of a guinea pig, inoculated from a case of tubercular pleurisy in the human subject.

The growth on the serum progressed slowly and took the form of a thin, white film in which were sparsely scattered, thicker, more opaque areas.

Stained with carbol fuchsine warmed on the cover-glass, and treated with 25 per cent. sulphuric acid in a watch-glass. If used in the cold, the dye should be allowed fifteen minutes.

The culture consists of straight and slightly-curved rods with rounded ends, and of varying length, the shortest hardly more than oval. The bacilli tend to occur in small parallel groups arranged in irregular lines, or set at various angles to one another in a way suggestive of Chinese characters. They exhibit none of the segmentation, or "beading," so commonly presented in phthisical sputum (see Plate XXX in the previous volume).

Branching Forms.—On the right are shown, from the same culture, two examples of the filamentous forms, which recall the branched mycelium of the hyphal-fungi or moulds.

The colouration of the bacilli (stained by Gram's method) is almost limited to the minute spherical granules lying within the bacillary cell.

Branching forms are at times met with in phthisical sputum.

Plate XXII.

Fig. 1.—BACILLUS OF RHINOSCLEROMA.

From a streak culture on agar, of forty-eight hours' growth, incubated at 37° C.

Carbol fuchsine, washed in dilute acetic acid. The figure shows a group of bacilli united into a zoögloea by an abundant ground substance which is faintly stained with the dye. The micro-organism consists of spherical elements occurring singly, but most frequently in pairs or short chains. Rod forms may also be met with.

As showing the essential similarity of ground-substance, or glœa, and that which constitutes a bacterial capsule, it will be noticed that whilst at the periphery, where active multiplication is proceeding, the bacilli lie closely embedded in this substance, more centrally, where the latter has accumulated in larger amount, it has parted into areas appertaining to definite groups, showing the divided share which the organisms have taken in its production. At the lower end of the figure, near its middle, there is a diplo-bacillus quite isolated from the neighbouring mass, and furnished with a capsule proper to itself.

Fig. 2.—STREPTOTHRIX ACTINOMYCES.

Broth culture, of seven days' age, incubated at 37° C.

In broth the growth occurs as small spherical colonies, which are best examined by being teased out on a cover-glass after having been washed in distilled water. The preparation is allowed to dry and may be then stained with carbol fuchsine or aqueous solution of gentian violet ; in the former case the subsequent washing is carried out with acidulated water, in the latter with tap water, or. what is better, with a 1 in 10,000 solution of caustic potash, followed by tap water.

The culture consists solely of long, interlacing, slightly wavy filaments which give off lateral shoots. The branches vary in length according to their age, taking, on their first appearance, the form of minute excrescences.

Cultures of the streptothrix Maduræ (the cause of mycetoma or Madura disease) are indistinguishable in microscopic features from streptothrix actinomyces.

As found in the tissues of oxen affected with actinomycosis, the filaments. as a rule, terminate in clubbed enlargements, and radiate from a central mass. In the lesions of the human subject the clubs are by no means regularly present, nor is a radial disposition of the filaments always obvious.

Such differences possibly indicate that the streptothrix causing the disease known clinically as actinomycosis in man and the lower animals is not always identical, the existence of pathogenic varieties (which have received special names) having been demonstrated by means of artificial cultures.

Owing to its anomalies, the classification of this group of organisms has been long a subject of controversy, which has been terminated, at least for a while, by raising it into a distinct class of fungi under the name of streptothriciæ.

The group includes non-pathogenic as well as pathogenic forms. and comprises micro-organisms which, like hyphal-fungi or moulds, form a mycelium of branching filaments originating from spherical spores ; certain of the filaments (like the aërial hyphæ of moulds) rise into the air from the mycelium and produce at their extremities chains of spherical elements comparable to spores, though of a different morphological nature from, *e.g.*, the endo-spores of bacilli.

PLATE XXII.

Fig. 1.—Bacillus of Rhinoscleroma.

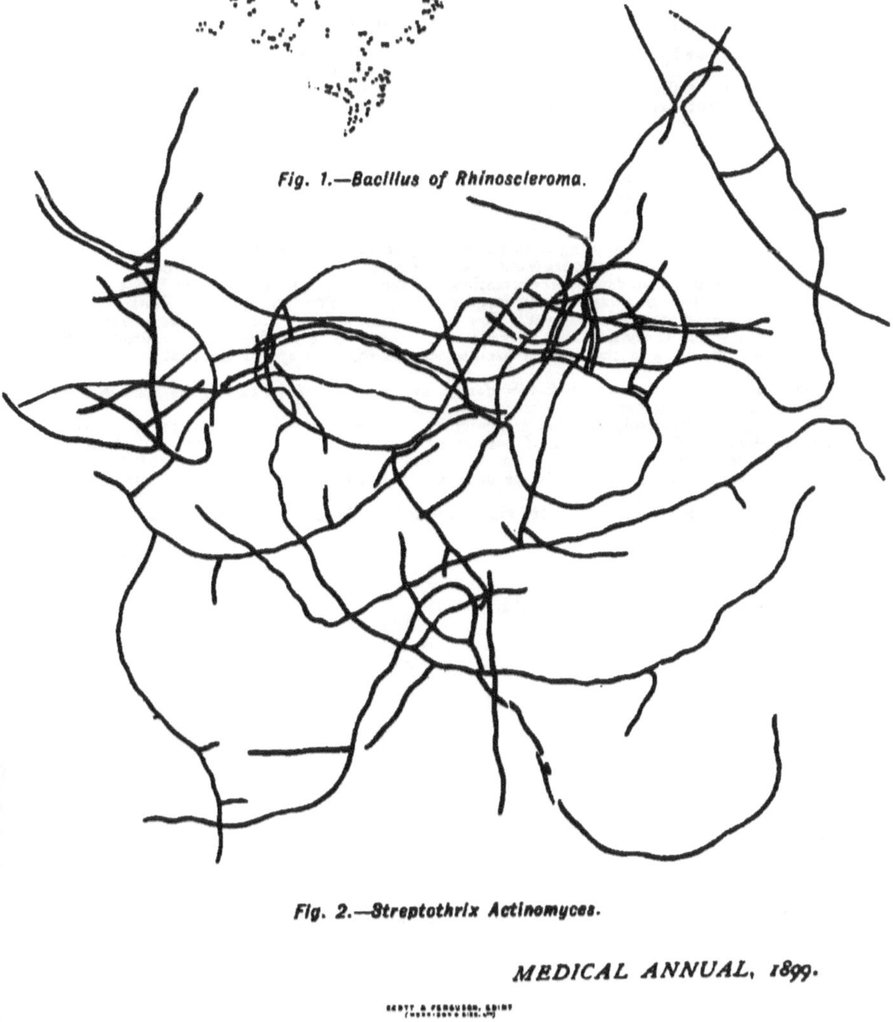

Fig. 2.—Streptothrix Actinomyces.

SCOTT & FERGUSON, PRINT

PLATE XXIII.

Fig. 1.—STREPTOCOCCUS PYOGENES.

Broth culture of forty-eight hours' growth, incubated at 37° C., from the pus of one of the subcutaneous abscesses which arose during the course of puerperal septicæmia.

Carbol fuchsine, washed in dilute acetic acid. The chain on the right hand side of the figure is added from a cover-glass preparation of the pus of an axillary abscess which showed large numbers of such without any staphylococci. The micro-organism, which is best studied in broth cultures, presents itself as micrococci arranged in chains of varying length. The component elements are almost everywhere flattened, and occur in pairs, showing that active division is in progress throughout the chain. Here and there elements occur which are oval or elongated in the direction of the chain, and all transitions may be traced between such and the pairs of flattened cocci resulting from their sub-division. Wedge-shaped forms may be met with in adaptation to the pressure arising at sudden bends. At times division of a component co cus takes place in the long axis of the chain, as may be seen in that on the right hand side. This mode of fission may become a source of lateral branching should the process of sub-division continue in parallel planes.

At times certain of the cocci will divide cross-wise into four. Some of the chains are very slender, and different parts of the same chain may present marked differences in respect of breadth or thickness.

Fig. 2.—BACILLUS OF QUARTER-EVIL, OR SYMPTOMATIC ANTHRAX.

Cultivation in 2 per cent. glucose broth, incubated at 37° C., of twenty-four hours' growth.

The organism, like those of malignant œdema and tetanus, is a strict anaërobe, i e., it grows only in the absence of gaseous oxygen. The culture was grown in an atmosphere of nitrogen by Buchner's method of removing the atmospheric oxygen with pyrogallate of potassium. Carbol fuchsine, washed in water weakly acidulated with acetic acid.

The growth at this stage consists of rods of varying length produced either into simple or segmented filaments. The elements composing the filaments are by no means of regular length, in consequence of a continuance of their sub-division.

The breadth is less than that of the anthrax bacillus, and the apposed ends of the segments, in place of being squarely cut (as in the latter; see preceding volume), are rounded.

It must be here observed, however, that in the bacillus of anthrax squareness of the apposed ends may be as wanting as in either malignant œdema or quarter-evil.

At the end of twenty-four hours few sporulating rods were present in the culture.

By the third day abund:nt spore-formation had taken place, as is shown in Fig. 3. The preparation consists of straight rods, large numbers of which are sporing. Pairs of rods joined end-to-end are not infrequent. The spore forms, as a rule, at one of the extremities of the rod, which it considerably exceeds in diameter so as to give rise to a drum-stick appearance, much as in the tetanus bacillus, except that the spores are oval in place of being spherical.

The formation of the spore is first evidenced by a swelling of the end of the rod : in this enlargement the unstained spore subsequently appears.

This organism is closely like that of malignant œdema both in its cultural characters and its morphology. In the bacillus of malignant œdema, however, the spores, in place of being terminal, form towards the centre of the rods. The bacillus of quarter-evil has not yet been identified as a cause of disease in man; that of malignant œdema not infrequently has.

Stain : Aqueous solution of gentian violet, washed in 1 in 10,000 solution of caustic potash, followed by tap-water.

PLATE XXIII

Fig. 1.—*Streptococcus Pyogenes.*

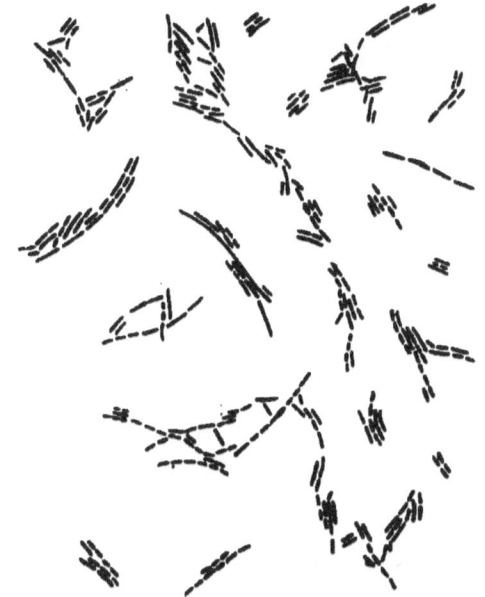

Fig. 2.—*Bacillus of Quarter Evil.*

Fig. 3.—*Bacillus of Quarter Evil
(sporing stage).*

MEDICAL ANNUAL, 1899.

SCOTT & FERGUSON, EDIN?
(H200-300-2 R.30. JR)

PLATE XXIV

FIG. 1.—DIPLOCOCCUS PNEUMONIÆ.

Culture on Löffler's blood-serum, of twenty-four hours' age, incubated at 37° C.

Stained by Gram's method. The cover-glasses having been prepared in the usual way, are immersed in a watch-glass of aniline gentian violet for ten minutes, passed through water, and then placed in Gram's iodine solution for five minutes, after which they are washed in alcohol until no further colour comes away ; they are then placed on edge to dry, and finally mounted in xylol balsam.

The cocci occur singly and in pairs, but mostly grouped in rows of varying length. In the particular strain shown (which was isolated by Dr. J. W. Washbourn and Dr. J. Eyre, and found highly virulent when tested upon animals) the chain formation is unusually pronounced, the organism being a streptococcal variant.

Considerably longer chains than the longest depicted were present. The cocci composing the chains are actively subdividing, as evidenced by the flattened pairs of which they so largely consist. Here and there unusually large elements occur in which sub-division has not yet taken place.

In pure cultures the organism is unprovided with the capsule which it presents when studied in the sputum and pulmonary tissue in cases of acute pneumonia.

FIG. 2.

A preparation of the heart-blood of a rabbit experimentally infected with the foregoing strain of diplococcus pneumoniæ by intra-peritoneal injection. All the organisms, whether single cocci or pairs, are surrounded with a thick capsule. The specimen was dyed with the following modification of dahlia stain devised by Dr. A. MacConkey: Dahlia, ·5 grms. ; methyl green (oo crystal), 1·5 grms. ; saturated alcoholic solution of fuchsine, 10 c.cm. ; distilled water to 200 c.cm.

The dahlia and methyl green are rubbed up in a mortar with part of the water until dissolved, the fuchsine is then added. and, finally, the rest of the water. The cover-glass, after the dye is placed upon it, is held over a low flame until the steam rises, placed aside for five minutes, washed, allowed to dry, and, finally, mounted in xylol balsam.

The appearance of the capsule under the conditions of the experiment and in the human tissues possibly marks a defensive formation on the part of the bacterium to protect it against the action of the cells and body fluids (Louis Jenner).

FIG. 3.—TETANUS BACILLUS.

Culture in glucose broth, grown anaërobically at 37° C. ; stained with gentian violet.

Slender rods and simple filaments. Most of the rods have sporulated, the spore being of spherical form and situated at one end of the rod which, in consequence, acquires the appearance of a drum-stick. Branching filaments may be met with, as in the case of the tubercle bacillus.

PLATE XXIV.

Fig. 1.—Diplococcus Pneumoniæ (pure culture, encapsulated).

Fig. 2.—Diplococcus Pneumoniæ (capsulated condition).

Fig. 3.—Bacillus of Tetanus.

SCOTT & FERGUSON, EDIN.

PLATE XXV.

FIG. 1.—BACILLUS PNEUMONIÆ (FRIEDLÄNDER).

From a streak culture on agar, of twenty-four hours' growth, incubated at 37°C.

Stained with aqueous solution of gentian violet, washed in caustic potash solution (1 part to 10,000), followed by tap-water.

Thick rods of varying length, the shortest appearing as ovals or even as coccus forms or spheres. Pairs of oval (or spherical) elements, end-to-end are not uncommon. In some of the shorter forms one or more circular vacuoles are present.

In pure cultures the micro-organism has no capsule such as it presents when found in the sputum or lung in acute pneumonia

It is not stained by Gram's method, and in this contrasts with the diplococcus pneumoniæ which is found in other cases of acute pneumonia, in the sputum and pulmonary tissue, etc.

FIG. 2.

SPIRILLUM OF ASIATIC CHOLERA (KOCH'S "COMMA BACILLUS").

From a streak culture on agar, of twenty-four hours' growth, incubated at 37°C.

Aqueous solution of gentian violet, washed in 1 in 10,000 caustic potash solution, followed by tap-water.

A group of typical, slightly-curved rods is shown at the lowest part of the figure; others are selected to show varieties of form. The rods vary in length as well as in thickness. Some are sharply curved like the letter C; others are curved in opposite directions and present two bends like a shallow S.

Some of the bacilli exhibit one or two circular vacuoles; the vacuole may occupy one end of the rod, which may be slightly enlarged. Although commonly rounded, the extremities of the bacilli may be bluntly pointed. In addition to the simple, curved rods, there are longer filamentous, undulatory forms of different lengths; these present no traces of a segmented or composite structure. The proper spirillar or twisted feature of the filamentous forms is appreciable only in a hanging drop made, e.g., from a broth culture. In such a drop even the longest forms are twisted, as appears by the alternation of parts within and out of focus. In dried preparations the curves are reduced to a single plane, the filament becoming undulatory or serpentine without spiral twist.

PLATE XXV.

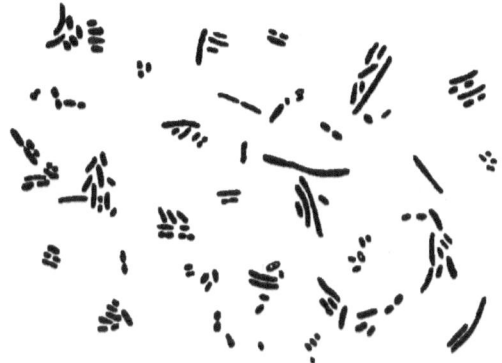

*Fig. 1.—Bacillus Pneumoniæ.—*FRIEDLANDER.

Fig. 2.—Spirillum of Asiatic Cholera ("Comma Bacillus").

SCOTT & FERGUSON, EDINR
[THOMPSON & CIE. LTD.]

is carried out as follows: The best culture medium is Löffler's blood serum, though ascitic or pleuritic, or even hydrocele fluid answers sufficiently well; the medium is solidified by heat in test tubes placed on the slant to furnish a surface on which the microbes may be sown. The serum slant may be inoculated by means of the platinum öse ; this having been first sterilized in the flame is drawn over the tonsil or pharynx (over the membranous exudation if such is present) and is then rubbed over the whole surface of the medium so as to distribute the organisms present as widely as possible. Or the serum may be more effectively inoculated by means of a small swab of cotton. The swabs for this purpose are made by passing a strip of cotton through an eye at the end of a piece of stout copper wire around which the material is then twisted for a short distance

Fig. 6.—A reduced figure of the swab, etc., described in the text.

(*Fig.* 6). The opposite end of the wire is bent to afford a hold, and a second piece of cotton is wrapped around near this as a plug which closely fits the mouth of the test tube, into which the swab is inserted.

A set of such tubes may be sterilized by being heated for an hour in the hot-air sterilizer at 150° C, and can be kept ready for use. When about to be employed the swab is removed from the tube and applied to the most promising area of the throat ; it is then introduced into the culture tube and gently moved over the surface of the serum. The swab may then be destroyed in the flame. After inoculation the culture tube is incubated at 37° C, having been first capped to prevent any drying of the medium. Incubation of the tube is essential, and without an incubator the further investigation is preferably handed over to an' expert or to one of the Associations or Institutes where such work is carried out.

Culture tubes and swabs may be obtained, moreover, from many such sources. After twenty-four hours' incubation the growth of the diphtheria bacillus appears as small hemispherical, greyish, shining colonies; where such have arisen so closely as to coalesce the circular outline is wanting. As a rule, colonies of other bacteria develop concurrently, such, e. g., as staphylococcus pyogenes aureus, staphylococcus pyogenes albus, streptococcus pyogenes, sarcina lutea, or forms of yeasts. The colonies of the diphtheria bacillus cannot be distinguished with any certainty by their macroscopic characters, though yellow colonies may be ignored. If no suspicious colonies appear for individual examination, the öse is swept over the culture and the material transferred to cover-glasses, on each of which a drop of distilled water has been previously placed, the examination being made in the dried state, by the technique fully described in Part I.

Labor may be saved by using a single, long cover-glass on which a series of such preparations may be made and stained, in place of separate circles or squares.

The films having been allowed to dry, they may be stained with carbol fuchsine or by Gram's method, but as satisfactorily and simply with Löffler's methylene blue as with any other dye:

| Concentrated Alcoholic solution of Methylene Blue 30 parts | Solution of Potash (1-10,000) 100 parts |

The stain is allowed to act for five minutes and washed off

in tap water, the cover-glasses being then placed on edge to dry or gently pressed between filter paper; when quite dry they are mounted in xylol balsam.

The diphtheria bacillus is depicted and described on *Plate XIX.*

If no diphtheria bacilli are found, a second series of preparations should be made from another öse of the same culture, and if a negative result is again obtained, a third set. Where any doubt enters into the morphological diagnosis it is imperative to test the virulence of the bacillus upon animals. With this object a tube of broth is inoculated from a colony of the bacillus, and after forty hours' incubation 2 cubic centiméters of the shaken culture are injected into the subcutaneous tissue of the anterior or lateral abdominal wall of a guinea pig weighing 500 grams. The broth used for the culture is prepared from minced veal which is allowed to ferment in order to remove the glucose present in it, and by so doing to reduce the amount of acid formed by the bacillus, the production of which acts deleteriously upon the micro-organism and diminishes the amount of toxin elaborated by it.

It must be borne in mind that diphtheria bacilli present all grades of virulence, the virulence at one end of the scale diminishing until it entirely vanishes.

There are, that is to say, diphtheria bacilli of typical morphological form which have, presumably, lost their virulence, and the injection of a broth culture of which produces neither local nor general results in the animal upon which they are tested. In the lower grades of virulence a local swelling only results; in the higher, death ensues in from twenty-four to forty-eight hours; in the case of very high degrees of virulence, much smaller doses of such a broth culture are lethal within the same time; death may be deferred for seven or even twenty-one days when the virulence is low.

The crucial test, however, as to whether a bacillus is truly a diphtheritic one is not, strictly speaking, its possession in general of a pathogenic property, but of one so specific that the local or general action of a broth culture is inhibited by the previous injection of the animal with anti-diphtheritic serum.

The true diphtheria bacilli, again, are characterized physiologically by producing an acid reaction in a 1 per cent. glucose broth.   This reaction is not given, e.g., by Hofmann's bacillus— a form sometimes present along with that of true diphtheria, or met with in cases of sore throat where none of the true forms occur (see *Plate*).

Lastly, it is a fact, with much practical bearing, that bacilli having the morphological characters of those of diphtheria, and highly virulent, as tested upon the guinea pig, may be isolated from the throats of persons, who themselves exhibit no disease and have not suffered from diphtheria; since such individuals, whilst themselves immune, may serve as carriers of infection. This has been particularly shown in the case of outbreaks of diphtheria in schools.

And, what is equally important, after a diphtheritic patient has quite recovered from the disease, that is to say, after having acquired an immunity from the disease by reason of having had it, he may bear about diphtheria bacilli in the throat for months, harmless enough to himself, but capable of conveying the disease to others.   It becomes, hence, strictly necessary to make repeated bacteriological examinations of the throat after convalescence is established; and if the patient is to be no longer a source of danger to others, his isolation may be maintained so long as the bacilli persist.

Formulæ of the stains referred to in this volume:

Carbol fuchsin (Neelsen's solution):

Fuchsin 1 gram | Aqueous solution of Carbolic
Absolute Alcohol | Acid (5 per cent.) 100 c.c.
10 cubic centimeters (c.c.) |

Aqueous solution of methylene blue:

Methylene Blue 2 grams | Distilled Water 85 c.c.
Absolute Alcohol 15 c.c. |

Löffler's methylene blue:

Concentrated alcoholic solu- | Solution of Caustic Potash in
tion of Methylene Blue | distilled water (1 in 10,000)
30 c.c. | 100 c.c.

Aqueous solution of gentian violet:

Gentian Violet 2.25 gram | Distilled Water ·100 c.c.

Or the following may be used, except for the coloration of the hanging drop (see p. 28).

Gentian Violet 1 gram | Distilled Water 80 c.c.
Absolute Alcohol 20 c.c. |

As gentian violet is a basic dye, the washing of cover-glasses after staining with an aqueous solution of this re-agent is best carried out in tap water, which is naturally slightly alkaline; distilled water extracts more of the dye, and acidulated water yet more. Better than tap water, however, is a solution of caustic potash in distilled water, one in 10,000, the use of which for this purpose was devised by Mr. Arthur Mead. The cover-glasses, after staining, are passed through two saucers of the potash solution, and finally through tap water.

A series of comparative trials has shown that the violet is in this way rendered more intense, and the result of the stain more certain. Or the potash, as suggested by the author, may be added to the dye. The following formula of Mr. Mead's gives excellent results, the specimens being washed in tap water after the use of the stain.

Potassic gentian violet :

Gentian Violet . 1 gram	Caustic Potash Solution in dis-
Absolute Alcohol 20 c.c.	tilled water (1 in 10,000)
	100 c.c.

Concentrated alcoholic solution of gentian violet :

Gentian Violet   25 gram | Absolute Alcohol   100 c.c.

Gram's method : *

(1) *Aniline gentian violet :*

Concentrated alcoholic solu-	Aniline Water 1,000 c.c.
tion of Gentian Violet 11 c.c.	

The solution is to be freshly made, and filtered before use. Aniline water is prepared by well shaking 4 c.c. of pure aniline with 100 c.c. of distilled water, and twice filtering through paper, first moistened with distilled water.

(2) *Iodine solution :*

Iodine 1 gram	Distilled Water 300 c.c.
Iodide of Potassium 2 "	

The preparation of distilled water for the purpose of making cover-glass films (described on p. 26), may be expeditiously carried out by allowing the steam of a beaker or kettle to condense on the clean under side of a flat capsule, partially filled with cold water, and afterward quickly inverting the latter, when ample will be found on the side brought uppermost.

* [The thin smear is dried on the cover-glass, fixed by heat, stained with aniline gentian violet for five minutes, immersed in the iodine solution one or two minutes, washed in strong alcohol until nearly colorless, dried and mounted with balsam.

www.ingramcontent.com/pod-product-compliance
Lightning Source LLC
Chambersburg PA
CBHW032355020726
47499CB00008B/2752